Our 12-Dimensional Universe

By Steve Preston

1st Edition

Table of Contents

Introduction

In 1967, blue flames in a window brought the fire department. What they found was the body of Robert Francis Baily with bluish flames still coming out of his abdomen. Hundreds of other cases of spontaneous human combustion make us wonder about reality.

We read in our Bible Elisha, Elijah, Peter, Paul and others brought people back to life and we question was is real.

Recently, Canadian scientist, John Hutchison blasted a bowling ball with ultrahigh beat frequencies of microwave transmitters and it levitated off the ground.

Ancient texts confirm that The Jewish prophet Moses, Egyptian magician Jambres, and Ancient Mystic Mater all converted sticks into snakes and had them turn back to sticks. Could this be possible?

Today, astronomers tell us the red shift calculations of a multitude of stars pinpoint the location of the Big Bang where the Earth is today, so could the Big Bang have happened?

A multitude of people of people reportedly cured blindness and other diseases by touch from Biblical times and into the present time. Many other things that appear to be miraculous occur and no one seems to describe how it happens.

Idiot savants listen to a piano and automatically know how to and play the piano with ease as if they had played the piano before they were alive.

In 2016, physicists at Delft University in the Netherlands, instantaneously transferred quantum data concerning the spin state of an electron to another electron about 10-feet away to repeat other experiments showing the space-time barrier does not exist "all the time".

A number of young adults were able to hoist automobiles off their fathers and loved ones to save them without their bones crushing, without ligaments being twisted and with no other damaging effects as if the automobile was suddenly lighter for an instant.

Massive particles of the universe appear to be stable and clumped into galaxial groups totally against the law of Entropy. Massive Galaxy walls are being found containing dozens of massive galaxies. How can this possibly happen?

On and on we could go with unusual events. They all have one thing in common. Either the physical properties of our universe require them to happen or they are lies. What I will show you is that these certainly can happen according to the dimensional characteristics of our universe. Please do not discount unusual events, and astonishing characteristics of our great universe by thinking there are only 4 stupid dimensions that characterize our universe. No wonder things don't make sense. You were taught that height, length, width, and time made up everything. Today we know that time can completely disappear just from traveling fast and space can turn to nothing. By the very stupid 4 dimensional characterization, Einstein's relativity, religion, and Quantum Mechanics are automatically lies. This is no way to teach children about reality. All it does is destroy any chance they have at understanding life.

Originally, I wrote a book entitled "Our 10 Dimensional Universe". Everything seemed fine as I developed what I thought was an understandable description of our universe before I realized how important the characterization of Participatory Anthropics was to our quantum fluctuating universe and our understanding or it. For those not fully aware of what I'm talking about, let me bring up Schrodinger's Cat. It seems like everyone has heard a joke or 2 about this thought experiment. Dr. Schrodinger suggested that if a cat was placed in a box and there was opportunity for it to be killed with poison in the box that, until someone actually opens the box, the cat **is** both alive and dead. It does sound like a joke until you actually look at the universe. When Einstein was asked the question *"If a tree falls in the woods and no one is around does it make a sound?"*; he simply said. *"There is no tree."* More and more people now know there is no tree if there is not a cognizant viewer of the reality associated with a tree.

Our reality is not <u>valid</u> without us.

I know all of this sounds like I'm talking in circles but I think you will gain a much better perspective of our reality after understanding the 12-dimensions by which we can define it. If you already know what they are and how they work, there is no need for you to read this book. If you are somehow satisfied that there are only 4 dimensions with length, width, and depth being dimensions of reality, I feel sorry for you and you don't need to read the book either. This book is for those who want to understand how a little girl can lift a car off her father and save his life, when her height, length, and width tell us it cannot happen. For those wondering how scientists can tell us all stars are going away from us such that the Earth is the center of the universe when the Big Bang Theory tells us there is nothing at the center of the universe, the book will give answers. If you ever wondered if time travel was possible, this book will provide answers. If you wondered what happens when you die, this has some answers. For a religious flavor, Jesus told his disciples that with faith they could change the characteristics of mountains or walk on water. We will find that is certainly is possible, so don't go putting down the Bible just yet. There may be important things in it that will help us understand ourselves and what reality really is.

While I'm on a role, let me bring up one more Bible thing. Genesis 1:1 says that "light" came well before the Sun was seen in the heavens. This sounds like a gross error in a three dimensional, non-relativistic, non-anthropic universe and many scoffed. Now we are finding out that it has unbelievable insight into the way reality can be presented without upsetting what we call physics. Here is the thing. If there are anomalies in what you see, have read, or have thought about, and they don't fit into a stupid universe made with an inanimate width, length, and depth, you need to open up to another possibility that allows for living creatures with awareness to survive. Our universe and its supersymmetric adjoined universe allow for life and allows for much more. All we need to do is to understand it a little better. Before we get started let me just say what the 12 dimension that create our universe are. Let me warn you right now they will sound weird until we discuss them a little.

- *There are 3 perpendicularly recognized dimensions that build particles matter and even black holes.* [This is a little bit like the length, width, depth stuff, but it truly involves so much more.]

- *There are 3 perpendicularly recognized dimensions that build forces that hold matter together, tear it apart, and*

8

cause something we define as light. Some of the nuances of these dimensions are Electricity, Electromagnetic fields, and magnetism. [We actually know more about these dimensional qualities of our universe than the others simply because we believe we can see light.]

- *The third set of perpendicularly recognized dimensions builds life itself including self, the soul, and the spirit of life.* [Please, please, please understand that reality is only relevant when a cognizant observers is aware of it. I'm not just talking about Schrodinger's cat, I'm talking about you and me.]

- *This leaves a final 3 perpendicularly recognized dimensions of space-time, including forward and backward time, the appearance of space, and the space-time continuum.* [I know you would like to think of time being a standard, but good old Einstein tore that to bits when relativity entered our knowledge base.]

I know you are saying those aren't dimensions; they are just bizarre notions, but today, we know that going the speed of light halts time completely and forces mass to infinity. We know that particles can be transported instantly without going through space and we know there is no such thing as light as vibrations of nothingness enter our eyes and our brain somehow converts the vibrations to something entirely different. We also know that as light vibrations speed up, light changes to something very dangerous that can destroy life. Today experimenters have made things become invisible, passed things through metals without affecting the metal, and all sorts of strange things that simply don't fit into the simple 4 dimensional absurdity you were taught in school. Let me give you a very brief history of dimensional discovery.

Dimensions are Discovered

The concept of a simple universe started changing as John Keely described everything as being made from vibrations of an unseen no mass sub-particle he defined as <u>Aether</u>. Einstein used this Aether as the basis for all matter. It no longer was true matter. Instead it was simply vibrating nothingness. The faster it vibrated the larger the apparent mass. Unfortunately, the waves of vibration had no end so matter got larger and larger and less dense and less dense until it reached the "End of the Universe" where it disappeared forever. As one cannot create matter, our universe would slowly be destroyed by loss of these vibrations. Light did the same thing. It continued to the end of the universe and was lost forever. Soon, Einstein thought all light would be gone from the universe.

Dr. Milo Wolff came to the rescue as Einstein became worried sick. He recognized that our universe only worked by having a symbiotic union with a "coexistent linked" universe. Dr. Wolff considered these mass vibrations as they left our universe [what he called out-waves]. He surmised that the linked universe saw them as waves coming into their universe. He called these things in-waves. As the in-waves excited vibrations in the adjoining universe, they emitted out-waves [Mass] which left that universe and came into our universe as----you guess it----in-waves. Possibly you can tell from this, but what is being said here is that time goes backwards in our joined universe. This assures we will never run out of time and it will always be pushed back into our universe as time goes to infinity and leaves out universe.

11

Please understand, I know this all seems bizarre to you because you were not taught it in school, but I assure you it will make sense later on. Our joined universe allows us to have conservation of time.

Conservation of Energy LAW

Eureka! The method that assured mass energy and time could not be lost was modeled and everyone sighed! Mass energy and the perception of time are always the same because of this rejuvenation. No matter what we did, there would be a counter action to restore everything using something defined as Super-Symmetry. If we make something here, there is the opposite happening in this linked universe; therefore we can change something here very easily. If a fusion reaction builds larger mass atoms in this universe, there would be an equal reducing of atom mass to regulate the system and allow it to happen.

In-Waves

Let me just talk about out-waves [essence of matter] and the in-waves [essence of forces that allow matter to do work or change a static environment.] Without in-waves, there could not be out-waves and vice-versa. This is the crucial description of dimensions. This will make more sense as we go along. Don't worry if this is confusion, right now, it will make more sense as we go along. Right now let me introduce you to a new way of looking a matter as these vibrating "Out-Waves". Actually, what happens is In-waves come into our universe and intersect out-waves of vibrating Aether [remember I told you Aether was vibrating nothingness or quantum fluctuations, but it is still very important to us as everything is made from Aether---- including nothingness.

12

Let me put it this way; an atom is a construct of Aether being fluctuated as a quantum characteristic. At the same time, an atom is a standing wave at the intersection of in and out-waves as they proceed through our universe; and finally, an atom is an accumulation of similar vibrational attitudes of Aether. I know all three definitions sound like mash potatoes and you are thinking about throwing the book in the trash right now, but if you do you may never understand how reality works.

Pleased believe this will begin to make more sense as we go along---I promise! While you are already confused let me just bring up a few more things so that when we talk about them, they would be new to you.

Generation of Matter

It would be nice to simply say an atom is the simplest description of matter, but by doing that, we will not understand most of what our universe actually is. As externally generated vibrations called in-waves [a visual image of electro-magnetic force] strike internally generated out-waves [a visual representation of a vibrating Aether], they produce what Dr. Wolff called "standing waves" which takes on the characterization of matter. The faster the "standing waves" vibrate, the larger the "apparent" matter is as we find next.

Control of Time

Before we go on, I'll let you in on a secret about time. As our joined universe has backward time [called anti-time] and backwards matter [called anti-matter] and backwards everything, if we want to travel through time backwards, all we need do is to travel forwards in our joined universe.

13

Realities of Physics

While Physics holds true, all of the things you thought were impossible may not be.

- *Can you travel through time? Yes*
- *Can you pick up a car? Yes [just vibrate a little faster and it becomes easier.]*
- *Can you turn lead into gold? Certainly*
- *Can you walk across water? It has been shown in recent experiments.*
- *Can you change reality better or worse by some means? Absolutely*
- *Can you die? We will get into that as we go along, but the short answer is NO.*
- *Was Schrodinger's cat both alive and dead? That is a more difficult thing to describe, but in a way, this is true as well.*
- *Is there a Heaven Universe? Without it our universe would not exist.*

I know all this sounds fanciful, but today experimenters are getting us closer and closer to understanding how we affect our universe and how the universe can be manipulated. It is done with vibration.

Common Material Frequencies

Let me start by showing you a couple of tables of actual and/or theoretical frequency and wavelength standards of common elements known today along with other attributes of other characteristics of our universe. The material frequencies have been derived from the various groups investigating vibration reaction of structures/atoms. How would you like some particles vibrating at 60 exahertz? That vibration causes Gold, as you can see from the list following. Have the right frequency or resonate the environment around a substance and one can make ANY material you want. You can even vibrate yourself to modify the environment. Notice that most frequencies do not form matter, at least structures with mass. Even the smallest physical component [BOSON] must vibrate fairly fast [300 MHz] so one would think that if you wanted to modify particles, you had better have a source that can vibrate very, very fast. One thing to note as you look at the tables; vibrating frequencies that create the element we call Meitnerium can even vibrate faster to produce the limits of matter to what we call pure magnetism. Some call it a black hole. It is known that the event horizon of a black hole can take dark matter and convert it into much higher frequency "visible matter" and we find that we can do the same thing if we resonate at a high enough frequency.

15

Chart of Particle Vibrations

Name or characteristic	Maximum Wavelength [meters]	Highest Frequency [Hertz]
Aether [??]	*1 x10^{+10}	<30 x 10^{-3}
Fermion [part mass]	*1 x10^{+4}	30 x 10^3
Boson [smallest mass]	*1 x10^{-0}	30 x 10^7
Baryon [electron]	*1 x10^{-3}	30 x 10^{10}
Hydrogen/1	1 x10^{-9}	30 x 10^{16}
Berylium/9	1 x10^{-10}	30 x 10^{17}
Silicon/28	3.5 x10^{-11}	8.5 x 10^{18}
Zirconium/91	1 x10^{-11}	30 x 10^{18}
Gold/197	5 x10^{-12}	60 x 10^{18}
Meitnerium/270	3.7 x10^{-12}	27 x 10^{19}
Straight Gravity	smaller	higher

This table represents the form of the structural dimensions that we use in our reality, but what about the "in-waves"? They are characterized as electro-magnetic force as shown next.

Chart of Electro-Magnetic Vibrations

Name or characteristic	Maximum Wavelength [meters]	Highest Frequency [Hertz]
Electricity	5×10^{10}	$<30 \times 10^{-3}$
Brain function	5×10^{7}	6×10^{0} to 10^{1}
Human hearing	1×10^{4}	20×10^{3}
VHF [radio]	1×10^{0}	30×10^{7}
UHF [radio]	1×10^{-1}	30×10^{8}
SHF [radio]	1×10^{-2}	30×10^{9}
EHF [radio]	1×10^{-3}	30×10^{10}
Microwaves	2.5×10^{-4}	12×10^{12}
Infrared [light]	1×10^{-6}	30×10^{13}
Visible light	4×10^{-7}	75×10^{13}
X-rays	1×10^{-8}	30×10^{15}
Gamma Rays	1×10^{-9}	30×10^{16}
Magnetism	lower	higher

** It is highly likely that brain function frequencies are simply catalyst for much higher frequencies actually used by our brains to store thoughts and images.

The thing we know about electromagnetic frequencies is that less input energy is required for a particular action the higher the frequency of the action. It becomes easier to attain a purer, higher quality resonance and the force becomes greater until it is pure magnetism. We can model life the same way and look at the next chart.

Chart of Life Function

Name or characteristic	Maximum Wavelength [meters]	Highest Frequency [Hertz]
Molecular Interaction	5×10^{10}	$<30 \times 10^{-3}$
Unaware Life	1×10^{4}	30×10^{3}
Life Awareness	1×10^{0}	30×10^{7}
Survival	1×10^{-3}	30×10^{10}
Sex	1×10^{-9}	30×10^{16}
Need for Companionship	1×10^{-10}	30×10^{17}
Need to help others [Self Actualized]	3.5×10^{-11}	8.5×10^{18}
Selfless Love	5×10^{-12}	60×10^{18}
Universal Understanding	3.7×10^{-12}	27×10^{19}
Insight into the External World	smaller	Higher

I know this list sounds completely foreign to dimension and they look more like a pictorial reference of some Buddhist monk, but please stay with me as we will find as we head through Anthropic science that how one "realizes" his existence actually changes his existence and the existence of those around him. There is something to "Think and Grow Rich" and Self-Actualization, you have read about. As positive thinking is expanded, happiness will increase. There are many studies and right now just check out the chart as life becomes a controlling set of dimensions in our tenuous Universe. No matter how you sense it, increasing your vibrational resonance frequency, increases your power over your environment. At the very high frequencies, there is almost no need for the environment. Eliminating the lower frequency elements allows you to sustain the higher levels longer and better and that brings us to the time dimensions. For this one our frame of reference may need to be adjusted a little. Let me just introduce it here and explain in later as this will take more background.

Chart of Time Vibrations

Name or characteristic	Maximum Wavelength [meters]	Highest Frequency [Hertz]	
No life understanding	empty matter	No universal awareness	Normal Universe
Normal existence vibrations	Normal understanding	Universe Awareness	Normal Universe
Faster vibrations	Enhanced awareness	Positive control	Normal Universe
Relativistic motion	Relativistic arena	heightened understanding	Normal Universe
Between life	Transverse light	Light speed	
Inverted Relativistic motion	Inverse-Relativistic arena	heightened understanding	Linked Universe
Fast backwards motion	Inverse awareness	Positive control	Linked Universe
Negative Normal existence	Normal Inverse understanding	Joined Universe Awareness	Linked Universe
No negative life understanding	Empty matter	No universal awareness	Linked Universe

Beat Frequencies

That brings us to crystals and something called "Beat Frequencies". You have probably heard about crystals having some magical power and dismissed it as some type of belief destined to go along with astrology and extracts of poppy seeds. The new model of atomic structure and this whole concept of vibrating particles may give credence to the notion that crystals hold magic. If you start with a crystal of a homogeneous material that is locked in a covalent lattice structure, it will tend to vibrate at a very specific frequency when excited and the vibrations will continue for some time due to the resonance of the crystalline substrate. In other words, a crystal could cause a continuing vibration. A secondary vibration from a sound cue or other stimulus could very well produce a "beat frequency". This means the two frequencies will "ADD" in a special way. If you add a 1KHz signal with a 1.1KHz signal, they can beat together and make a 2.1 KHz frequency that will beat again and make a 3.1KHz signal that can Beat with the 2KHz signal to make 5.1 KHz and so on. If you start with really high frequencies, the beat frequencies get huge. If you want the ability to sense the higher frequencies, simply dampen the lower frequencies or ignore them. Here is the weird part that has been experimentally accomplished. Get a high enough frequency and a different material could be created.

Caution

Don't discount the magic crystal thing, but don't go out and get a crystal to make you feel better either. It probably will just sit there and do nothing for you. That will have the opposite effect as structural vibration levels require focus outside the normal environment.

Beat frequencies of Life

Just like the previous example, if 2 consciousnesses get together and somehow raise their frequencies, just being near another will make the frequencies even higher as they beat together. When one person is happy, the other person will feel better. If, on the other hand, a depressed person comes near, you will probably feel a little worse than normal. There have been hundreds of experiments on this theme and all are spooky. People can solve problems faster if many people are trying to solve the problem EVEN IF THEY ARE NOT COMMUNICATING WITH EACH OTHER.

Bad Beat Frequencies

This whole *"get in a group trying to expand awareness"* thing has just as bad of an effect. If a mob of angry people get together, they lower the consciousness frequencies of the entire group [Hate, Lust, Survival, etc.], the entire group 'Vibration" will be lowered EVEN IF THERE IS NO VERBAL CONTACT BETWEEN EACH OF THE MOB MEMBERS. This has been proven over and over again. This is because beat frequencies go both ways. A 1 KHz signal and a 1.1 KHz signal ALSO make a 0.1 KHz "SUBTRACTED" frequency. We see the beat frequencies in electro-magnetics all the time and we can easily sense beat frequencies of sound pressure. Believe me when I tell you, the same happens with consciousness.

Get with the wrong people and you can greatly limit any greatness you can have. Get near the right people and your world will be opened up. It's not magic it vibrations and resonance.

Ancient humans discovered that vibrations, beat frequencies and power were all related. Open your mind to possibilities that ancient humans could do marvelous things with crystals and other particles that we are only now beginning to understand, but there is the warning. Because they were talented in these exotic ways, the people of that time quit listening to the creator God and started having self-centered thinking. They were empowered by their own science and did not understand how the ethereal dimension had to react to allow for true happiness and peace. Many ancient texts talk about how horrible the science was during that time and how the people had no idea about what that science was doing to them. That being discounted, there were many discoveries associated with changing vibrational resonances of materials and electro-magnetics. Let's look at some of them. As we do try to compare these physical vibration enhancing things to what might have been possible if they enhanced their vibrations.

How Do We Proceed?

We start with vibrational matter, but what we find is that everything in the universe is "vibration", not just matter. While many already knew that electro-magnetism was an essence of vibration and that somehow light was an emanation of those vibrations, we might not be as aware of the increasing understanding that life is also some type of vibrational emanation. This book will dive deeper into just how much of our universe is controlled and built around vibrating nothingness.

A New Theory

If you have read my last book on vibrational matter, you may appreciate that all matter is made out of something we can call vibration. People searched for years trying to find the elusive "unified particle" that would be the building block of everything, but no one could ever find it. Einstein spent the last half of his life searching for a particle that would be the building block of everything. What we know today is that atoms are made up of something called Baryons like "electrons" and "protons". These baryons are not the building blocks of matter because scientists found things that make up the baryons called Bosons. These things were the smallest particles that could make up matter or could exist as an entity to be sure, but there were several types.

Soon scientists found out that something called a fermion was less than a particle. Oh! My! This must be the smallest particle of matter. A graviton, for instance, is not quite a particle. It doesn't conform to matter in that it displays gravity, but there is no mass to make the gravity that is sensed. Unfortunately, several types of fermions were discovered, in fact photons themselves with their unusual character of sometimes having mass and sometimes not having mass put them into the fermion class.

What in the world are we to do????? [I'm making tiny little cry sounds. They are fake so don't worry about me.]

Eureka!

The answer is both simple and almost impossible to understand. Everything---I mean everything, is made of vibrational nothings. This includes all matter, all electro-magnetics, all nuclear energy, all photons, and even all life forces. This book is about how these vibrations come together to build a universe. String theorists have told us about adjacent universes like the one we call heaven, but we must look closely at our own universe before additional ones can make sense. The concept presented here will help you understand yourself, your religion, your life and death, you entire universe. While we are at it, I had to add a few dimensions as were theorized by the M-Theory that is the premier string theory of the day. String theorists mathematically determined that the dimensions existed, but, until now, no one has presented what they are. I didn't want to define these things. I had to do it because life was making me so confused. Don't get comfortable trying to define life with a length, a width, and a depth and don't try to use those "normal dimensions" to define a photon. I tried it and it only makes you crazy. It can't be done, so people get confused at

24

life, death and everything in between. This book will relieve you of stress by providing the beginnings of understanding of our world. Our world is made up of 12 dimensions.

Epiphany

Before dementia gets you as well, let me ask a simple question. Can you name the 12-dimensions of our universe? Our modern science tells us that they are out there, but no one seems to want to define them. After reading this book you will see that our universe is not the simple length-height-depth-time universe you have been used to.

In fact, the whole concept of length, height, and depth being "different" dimensions is pretty silly. All three of the main dimensions are the same things with a different directional vector.

How could "life" be created with that? Don't tell me life is made up of 2 lengths and a width. How could electricity be generated? I know in my heart that 4 depths, 1 length and a width do not cause electricity. How could gravity be defined? How could light be created? With light sometimes acting like a wave and sometimes acting like a particle, the number of lengths and heights must change from time to time. No matter what you do, more and more things come up that simply cannot be defined with some spatial separation augmented by time. Something had to be done. Scientist decided to make a new word.

Ekpyrotic Membrane

As we looked at in the first book, to find out more about our universe we will use several current theories and quite a few new ones. The first 2 are the Ekpyrotic [also called the Big Splat Theory] and Membrane Theories [Together called the M-Theory]. Both are related and both are from a series of theories from something called "string physics". The Ekpyrotic Theory requires there to be at least 11-dimensional components in our universe. The Ekpyrotic addition indicates that there must have been at least 2 adjoined universes that splatted together to initiate what we call the Big Bang some 15 billion years ago [In one time reference] or the Big Bang simply could not have occurred. This SPLAT action caused everything to emerge from a 'fireball' with a temperature of 10 billion degrees. When the two splatted together, the energy that became matter was introduced. I'm not going to dwell on these things, but I think they are important enough to review before we get into some fairly strange discussions. According to this theory tiny quasi-particles called fermions were all over the place. I'm sorry, but you will have to remember this word, because quite a bit of our universe uses these fermions as the lowest basic unit that has "some" measurable characteristics.

Missing Matter

What I mean by quasi-particle is that these fermion things were missing components that they needed to exist in our universe or the other one. They weren't quite particles. Don't go laughing, and don't think that this is my theory. Almost all string theorists have this same concept. All I want to do is to define the dimensions for you and explain how dimensions can be understood much better when we define what particles are and they are not what we learned in school. Remember the old atoms and the periodic table and all that? They all exist, but they are not the components that define existence, they are sort of locked into place by something called quantum.

Splat

When the universes collided 15 billion years ago, the fermions got all twisted around and turned into complete particles we call Bosons. It is these bosons that make up most of what we call matter. The question is what actually caused the fermions to become matter and what things are necessary in this universe for sustainment?

The reason Aether became fermions, which became matter, is simple. They began to vibrate. They were not the only things vibrating either.

What we will find out in this book is that the entire universe is more easily defined as vibrational fields that the more common length, height, and width that has been forced into your head. Energy of everything is governed by the vibrational characteristics. As an example let's investigate voice boxes.

Voice Boxes

Women have shorter larynxes than men and the things vibrate at a higher frequency than most men's larynxes. Men must push much more air and with more pressure than women to get the things moving so women can speak twice as long as men while using up the same or even less energy. It is a fact of life. Women don't get tired talking on the phone. I decided that once I knew this little secret, I should be able to understand the vibrations that make up the universe but I was stopped by Einstein.

Einstein

This crazy guy was going around saying that the time-dimension energy equation was $E=MC^2$. I knew in my heart that he was wrong. After all, all energy equations are of the same form factor.

$$E= \frac{1}{2} KX^2 \text{ [universal law of potential energy]}$$

$$E= \frac{1}{2} LI^2 \text{ [universal law of magnetic energy]}$$

$$E= \frac{1}{2} CV^2 \text{ [universal law of capacitive energy]}$$

$$E= \frac{1}{2} MV^2 \text{ [universal kinetic energy equation]}$$

$$E= \frac{1}{2} I\phi^2 \text{ [Universal Inertial Energy equation]}$$

Where was the ½? No one seemed to care, but it bothered me just like a woman's ability to talk for many more hours than I could.

The more I tried to show Einstein was wrong the more it seemed to be right. It was as if ½ the mass in the universe was invisible and the invisible mass had to be "used in the equation". If you wonder what else is invisible, simply look at photons. This is really part 2 of the study of dimensions. As was determined from the last section, scientists agree that the universe is defined by at least 10-dimensions with "time" being one of the dimensions. Also we found out that the basic unit of matter was not the atom, and it wasn't even the unifying particle [BOSON], but instead, all things are made up of partial particles called fermions. Fermions got you somewhat confused so I brought out the fact that there were many types of fermions, so something even "smaller" was the combining element of our universe. The controlling element is vibration.

Logically, these dimensions grouped into static dimensions that produce matter, operational dimensions that can produce motion and work, and Ethereal dimensions that convert objects into living entities or produce life. Below are the general concepts brought out initially along with expansions that will hopefully make these concepts easy to recognize and use. If we know all about these dimensional elements, we may be able to extend our consciousness in and beyond this universe. While you had been comfortable with the 4-dimensional world you thought you lived in, there were massive problems. Let me give you some examples.

Realizations

You have this burning idea that gravity and *magnetism have a similarity, but both appear to be completely different. In fact, no one has ever truly defined what gravity really was in the first place.*

Gravity has nothing to do with length width, height or time, *but you're pretty sure that gravity does exist and it is REQUIRED for our universe to exist. A mass without gravity would certainly be odd.*

Sometimes Light is a wave and *sometimes it is a particle. I explained how it could be identified as a transfer from one universe to another, but a different answer might be easier to understand.*

The scientist guys came up with something they call *Nuclear force, but you can't understand it with 4-dimensional space.*

Magnetism and Electricity both work for us all the time, but *their existences are not covered in the normal universal definition. Certainly no one would try to indicate that either of those items did not exist.*

We also must consider life itself. The universe without life *might not exist. For us, we can certainly say it would not exist because we would not exist. Life is not DNA so don't go thinking it's covered by a volume of particles.*

How about our consciousness? *Does our consciousness exist? If it does exist the question might be "If I had no consciousness, would the universe cease to exist?" Some will say that people being here or not being here have no effect of the world at all. This type of thinking might not be completely correct and there is something we do know. This "consciousness thing" is not governed by the 4 normal dimensions. Therefore, we need to investigate how it is constructed in this universe.*

What about LIFE itself? *The question would be, "If there were no life, would the universe exist?"*

Luckily, in a vibrational world, we are not governed by the constraints of volume. Instead, a frequency domain world allows and attracts multitudes of volumetric entities as working units inside the frequency domain. We can leave time to its own devices [for now] and define a true universe. Let's try to find the dimensions. It's not hard, but you must think of the key dimensions as vibrating strings so that you can picture them. In a Time-Space perceived universe, it is quite easy to ignore key dimensions that are required to continue an existence [life and electricity, etc.] In a frequency based "truer" existence, some of these things are not as easy to ignore. In the first place, items defined as frequency based now become critical in our universe rather than simply being characterizations of our universe. Let me reintroduce the, now important dimensional elements.

A 12 Dimensional Reality

As I have been discussing, matter, without the forces needed to hold it together, is defined by its own dimensions or dimensional dynamo. I don't mean length, height, width, but the actual dimensions are mutually perpendicular in vibration similar to the ones we used to think meant

32

something. Length, height, and width only mean something when viewed from a single aspect with time and distance both as a constant which we now know is preposterous. These, still taught, fake dimensions have limited understanding of existence and life for a long, long time. The way to look at matter, as I just mentioned requires dimensions of Aether, Gravity and a union of the two to build a combining wave. Only when all are held together can something "exist" as a particle. Matter can't exist without energy to hold it together, but it is controlled by its own set of dimensions. It exists as a three dimensional dynamo consisting of electric potential, Magnetic field, and the electro-magnetic waves generated by their combination. Here are the Dimensional Dynamos so far.

Photonic Dimensions

- *Electrical Dimension [Potential for Light]*
- *Magnetic Dimension [Kinetic part of light]*
- *Electro-Magnetic Dimension [Perpendicular product of the others.]*

Structural dimensions

- *Aethereal Dimension [Potential for Matter]*
- *Gravitational Dimension [Kinetic part of Matter]*
- *Aetherio-Gravity Dimension [Perpendicular product of the others.]*

Life dimensions

- *Life or Self Dimension [Potential for life]*
- *Soul Dimension [Kinetic part of life]*
- *Spirit Dimension [Perpendicular product of the others.]*

Time

Here is where the last dimensional dynamo comes in. For those believing we would have a universe without time, I feel sad for you. Certainly, time can be defined by more dimensions. The idea that "life" doesn't age as quickly if one if going near the speed of light certainly shows there are more dimensions in time and the idea that God sees existence laterally [sees the beginning and end of time simultaneously] possibly does mean time can be expanded into additional dimensions as follows:

Time dimensions

- *Time Dimension [Potential for time]*
- *Inverse time Dimension [Kinetic part of time]*
- *Space-time Dimension [Perpendicular product of the others.]*

We can 'sort-of' perceive 4 mutually perpendicular dynamos---each with 3 mutually perpendicular dimensions of their own as shown below with space-time in the middle.

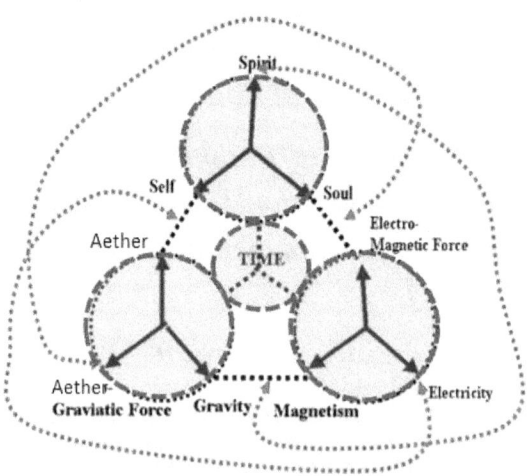

Let me slow down just a bit and tell you about each of these important dimensions.

Operational Group

As we know more about this group of dimensional characteristics than we do "matter" itself, I think we will start here.

Electrical Dimension- Electricity is the potential to create Electro-magnetic force. This dimension is characterized as a phase shifted, vibrating magnetic field. The lowest form of this dimension is called electrical or potential energy. Any time you see lowest form of a dimension in this book it means <u>a dimension with the lowest of "No" vibrational component.</u>

Magnetic Dimension- Magnetism is a characterization of the kinetic or vibrating electric fields. Actually, the magnetic fields are always perpendicular to electrical field. As you probably noticed all dimensional groups or Dynamos have similar characteristic definitions and actions, but the waves move differently in our universe.

Photonic Dimension What we call photons are actually vibrating quasi-particles of electro-magnetic fields. They are the electromagnetic interface to our linked universe. This allows the make-up of light to change drastically as frequency is modified by swapping energy with an adjacent universe.

All this energy swapping stuff will take some explaining.

Structural Group

Aether Dimension- Just like Electricity is the dimensional potential for electro-magnetics, Aether is the potential for building matter. From Aether the first perceivable characterization of matter is called Fermion. Fermions are the quasi-particles that vibrationally associate with others to form bosonic particles which build to Baryon particles which become Atomic particles, which become molecular particles which eventually make everything that is Matter. We can call the simplest fermionic component Aether if you like. Substantial amounts of our universe are made up of this simplest fermionic dimension component.

Gravitational Dimension- Gravity is a characterization of the kinetic or vibrating Aether. Gravity is perpendicularly associated with the force of matter and pure gravity is what we call a black hole.

Nucleatic Dimension- Nuclear attraction is actually vibrating interface to matter in our linked universe. It is this dimension that allows the creation of matter when matter cannot be created. I know you though Stephen Hawkins was crazy when he indicated that matter could be created at something he called the event horizon. Well this dimension allows that action by transferring matter in accordance with the law of super-symmetry. We think the most notable characterization of this dimension is called Mass force.

Life Group

Self- Dimension – We can say self is the potential for a life to become aware of reality. Life itself is made up of vibrating chemical responses set to some generally unknown master clock. It isn't the proteins of DNA that are life it is this vibration thing. Some call this dimension the "self" or self-awareness, or the EGO, but the thing to understand here is that there is a continuous struggle in one's life between this key dimensional quality and the other 2 life dimensions, what we can call the subconscious, and Spirit.

Subconscious or Soul Dimension- The soul is a characterization of the kinetic or vibrating self. The subconscious is not life but its vibrational essence is perpendicular to life. Some call this dimension the Soul of a person, or the ID. Some define this as the little voice in the back of your head or premonition or self-actualization characterization. This is going to sound weird, but think of this as a cause and effect because "the subconscious" reacts with "vibrations of life" like "magnetism" reacts to vibrating "electricity", or like gravity reacts to vibrations of matter. By affecting the vibrational level, "life" is changed and visa-versa. If a life vibration is halted, the subconscious must find another life to be able to sustain itself in this universe. I know I went spooky on you for a moment, but as you read this book please understand, I am not trying to devise a result. I'm simply trying to look at the stresses on the universe that make what we call reality. The later books will get into specifics associated with limited control of the universe by means of manipulation of the resonance of the three dimensional dynamos. Examples might be time travel, walking through a wall, levitation, converting matter into other types of matter. Like the electromagnetic dynamo this one also has a third component we can call the spirit.

Spirit Dimension- While the subconscious is very difficult to grasp initially, this "spirit" dimension is even more difficult, but just as important to our universe as it becomes the Self-soul force just like the dual of photonic and Nucleatic dimensions. Similar the others this is the "Key" to transferring life between this universe and our linked universe. The whole concept of angels appearing on earth as they transferred themselves for heaven is not a myth, nor does it violate some physical law. This strange dimension allows the creation of life just like the nucleatic dimension allows the creation of matter and the electro-magnetic force can create light.

Space-Time Group

The other 3 dimensions that characterize our universe can be characterized, tested, and understood from details written in ancient texts and from new sciences like Anthropics, String Theory, supersymmetry, and the like, but the new dynamo of dimensions should be the easiest one to reconcile especially after all the experiments showing time stops at the speed of light and quantum mechanics instantaneous travel without going through space, and people who have seen the future and come back to report on it. These things don't fit in our nice neat picture of time-space----because time-space dimensions help CREATE what we call reality. As they change so does our reality.

Time Dimension – Like Electricity, Aether, and the Self, the "time" dimension is the potential for a reality to be experienced in a time-space environment. If you go fast enough all understanding of time vanishes. If you could ever stop completely, time would last for all time. You would not have life, but there would be the potential for life. Some say you could see the beginning and end of times together. Time itself is made up of vibrating responses set to some generally unknown master clock.

Negative Time Dimension- The Negative Time dimension is not time but its vibrational essence is perpendicular and reversed from "normal" time. If you ever wondered what anti-mater is, one definition is that it is normal matter going backwards in time. One of the "jobs" of backwards time is to rejuvenate forward time. The second and most important "job" is to be the time-base in our adjoined universe. Lastly, it is reverse time that instantly converts matter into electro-magnetic forces to be used as the in-waves that drive our universe. Without reverse time, there would be no universe.

Space-time Dimension- The last dimension in the Time dimension dynamo essentially makes space-time. Like Black holes [Mass], monopoles [E-M forces] and spirits [Life], this dimension becomes a gateway to our linked universe so that the mechanism our universe can continue to operate for all eternity. One way to understand this dimension is something I call Lateral Time.

Lateral Time

In the space-time dimension, one can define an existence laterally where all time is instantaneous. Thought of as the state called the "speed of light", time halts and space goes to nothing in this unique construct. Beyond this speed time reverses and we are instantly in our linked universe. At this time we sense time backwards so one could, theoretically move in between universes and go anywhere in time. I hope you are following me, but if not, don't worry, we will expand on this to make it substantially clearer!

Symmetry Not Conservation

So you are told about Conservation of Energy, Conservation of matter with the push towards Entropy, and some may even told you about conservation of space-time and perhaps even conservation of life. All of these things seem to occur in our universe, but without an adjoined universe, just how is all this done? The answer comes from what I mentioned briefly as scientists study something call super symmetry.

As it turns out, symmetry is more important and more realistic than conservation of universal elements. You may know about the Theory of Conservation of Energy. Energy simply changes state rather than dissipates. It has an endless oscillation or vibration between Static and Kinetic elements. In a way, this is a true statement and certainly we can look at our universe in this way and things seem to fit together. What if I were to tell you that static energy, or what we call static energy is a characteristic of the out-waves from our universe and kinetic energy is a characteristic of in-waves from an outside universe. We don't conserve energy. It is continuously replenished. As our universe outputs out-waves, they are converted to in-waves and returned.

The universe dual MUST run symmetrically. The only way to apply force in this universe is from an outside force. AND Guess what!!! All dimensional qualities of this universe MUST BE symmetric with its universe dual.

41

Oh, you are a smart one!!! You are thinking, "That's not symmetry, the examples I keep bringing up show the opposite to symmetry.

Got You With This One!!!!

Remember that our linked universe has backward time. Therefore decreasing matter backward in time is exactly expanding matter in their time. Both universes would experience expanding masses as matter is created over here, anti-matter is increased over there in reverse time.

Backward Living

I've opened up a bag of worms now. If both universes sense time backwards to each other, there could never be interchange between universes------Right????-------

---

Wrong!!!!

In our linked universe, people experience time differently. It would be hard for us to understand how they see time, but I do think there is exchange between our universes so this oddness simply must be. I think that explanation will take one of those gurus on top of a mountain to explain. I'm having a hard time with forward time and I think I can still explain life without the extra confusion. Later, we will discuss the characteristics of the dimensions of time in more detail that may help.

Certainly, there could be more universes; the Membrane [M-theory] suggests many more universes could be co-resident with our "visible one" and so do many of the string theories and the symmetry of our universe may be shared, but it could also be indelibly linked and presented in this work. What I was saying is this. Matter, Energy, Life, Time and everything else in this universe stays in constant motion.

Never increasing and never decreasing without a secondary outside force.   Let me give some examples-

- If **kinetic energy** decreases, it increases in our joined universe which in-turn causes our STATIC Energy to Increase and vice-versa. [If we make a conversion of energy in our universe, the opposite WILL occur in the joined universe.]

- If **Gravitational energy** decreases, it increases in our joined universe which in-turn causes the exact opposite to increase in our universe. This increase makes us have an apparent reduction in mass so we will have to look at it a little. [If we force a reduction in gravity, our joined universe WILL experience an increase in magnetism to compensate.]

- If **Photonic [Electro-magnetic] energy** increases, it increases in our joined universe which in-turn causes the exact opposite to decrease in our universe [Because in-waves turn into out-waves, electromagnetism become particles in the adjacent universe so mass would reduce.]

- If **Internal Life energy** decreases, it increases in our joined universe which in-turn causes our External Life Energy to Increase and vice-versa. I know this whole external life energy sounds weird, so we will look at that a little as well. [The main thing is that as internal life here increases, it WILL decrease over there.]

- As Time vector goes in one direction, it MUST go in the exact opposite direction on a joined universe. **[Time must go backwards. This, of course is only from our point of view.]**

Certainly, these may not be the exact duals of our universe, but hopefully, it will give you an idea about how each one of

43

the elemental parts connect with both internal dual dimensions and external anti-dual dimensions.

To carry it one step further, we can establish that the energy bonds for the Particle dynamo are Static and centripetal forces. The energy bonds for the operational dynamo are associated with Kinetic and centrifugal forces. The opposite would be true in the joined universe. The anti-particle dynamo would be associated with Kinetic energy and the anti-force Dynamo would be associated with Static energy in that world. All this stuff is good information, but you really should be trying to learn some more about how to attain a higher level of awareness and a more meaningful life with respect to your life in this universe. Let me go over the strange statements in the Bible concerning this factor in dimensional variations in our universe.

Lose Life to Gain It

This section is more Biblical, but I think you can gain a lot by reading some of the very odd statements in this mighty book. They may not sound like information about dimensions at first, but I will explain after I bring out these points.

Mark 8:31-9:6- For <u>whoever wants to save his life will lose it</u>, but whoever loses his life for me and for the gospel will save it. <u>What good is it for a man to gain the whole world, yet forfeit his soul</u>? -- "I tell you the truth, some who are standing here <u>will not taste death</u> before they see the kingdom of God come with power." After six days Jesus -- led them up a high mountain, where they were all alone. There he was transfigured --And there appeared before them <u>Elijah and Moses</u>, who were talking with Jesus.

I mentioned before that saving your life to lose it sounds odd and the idea that one could or could not lose his soul perks up your ears, but then there is a very strange thing stated here. It sounds like the writer is saying someone "*will not taste death*" for 2000 years before God comes again in power. Finally, Elijah and Moses are both reincarnated to help bring light to what was being said. Matthew and Luke both indicate almost the exact thing, because this concept was so important.

Matthew 16:24-17- For <u>whoever wants to save their life will lose it, but whoever loses their life for me will find it. What good will it be for someone to gain the whole world, yet</u>

45

forfeit their soul?--- Or what can anyone give in exchange for their soul? For the <u>Son of Man is going to come in his Father's glory with his angels, and then he will reward each person</u> according to what they have done. "Truly I tell you, some who are standing here <u>will not taste death</u> before they see the Son of Man coming in his kingdom."-- After six days Jesus -- led them up a high mountain by themselves. There he was transfigured -- Just then there appeared before them <u>Moses and Elijah</u>, talking with Jesus.

***Luke 9:23-33**-For <u>whoever wants to save their life will lose it</u>, but whoever loses their life for me will save it. What good is it for someone to gain the whole world, and yet <u>forfeit their very self</u>? --- "Truly I tell you, some who are standing here <u>will not taste death</u> before they see the kingdom of God." About eight days after Jesus -- took Peter, John and James -- and went up onto a mountain to pray. As he was praying, the appearance of his face changed, and his clothes became as bright as a flash of lightning. Two men, <u>Moses and Elijah</u>, appeared in glorious splendor, talking with Jesus. They spoke about his departure, which he was about to bring to fulfillment at Jerusalem.*

Please notice one thing here, when God comes in power is defined as the time Jesus and his angels return and judge everyone for the things they did during their lives. Let me try to paraphrase if I can.

- *Jesus said if you want your "soul dimension to thrive", don't worry about your "the self-dimension or carnal life".*
- *He reiterated his command a second time indicating that if one <u>focuses</u> on success in this reality, he will lower the ability of his soul dimensional quality to control his environment.*

46

- *He reiterated it a third time saying there is nothing that we have as precious as our living soul-dimensional component.*
- *He reiterated it a 4th time saying some of the soul-dimensional elements of his followers will be reanimated in some way and still be alive [and conscious] when he returns in 2000 years.*
- *He emphasized a 5th time by having the living souls of Elijah and Moses come back into a human-like appearance to show these guys that their souls could not die."*

What a Contradiction!

Please notice that life and soul are sometimes interchangeable this is why he said saving your life loses your life. The second life is the life of the SOUL [most important part of you]. Let me give some more detail from John.

John 6:38-65-- *this is the will of him who sent me, that I shall lose none of all those he has given me, but raise them up at the last day.* *----The Spirit gives life [Soul]; the flesh[self] counts for nothing.*

Main Theme

We are told, at death, the "spirit" portion of a person leaves this universe and the Self/ Body turns to dust and dies away. That leaves the soul-dimension which can live "forever". This dimensional quality can be reincarnated even if the body has turned to dust. When separated from the Self/body dimension, the SOUL is, typically, in a completely unconscious state [still in control of developing our reality but unaware of time space to partake of its wonders. String theorists call this compactification of dimension. The Bible

47

indicates the soul dimension can be re-associated into this reality as shown with Moses, Elijah, Lazarus and other reanimations and a number of re-incarnations as new people.

*John 8:51-52--[Jesus said,] "If anyone keeps My word he [**his soul**] will never see death." The Jews said to Him, "— Abraham [**carnal body**] died, and the prophets also; and You say, "If anyone keeps My word, he [**soul again**] will never taste of death."*

*2 Corinthians 5:6-8--Therefore we are always confident, knowing that, while we [**our souls**] are at home in the body, we are absent from the Lord: For we walk by faith, not by sight. We [**our souls again**] are willing to be absent from the body, and to be present with the Lord.*

*1 Thessalonians 4:15--For this we say unto you by the word of the Lord that we [**soul part again**] that are alive, that are left unto the coming of the Lord, shall in no wise precede them [**souls**] that are fallen asleep. – the [**souls of the**] dead in Christ shall rise first; then we [**Soul part**] that are alive that are left, shall together with them be caught up in the clouds, to meet the Lord -So shall we [**our Soul with glorified, noncarnal body**] ever be with the Lord.*

*Matthew 16:28--Some who are standing here [**their souls**] will not taste death before they see the Son of Man coming in his kingdom.*

*Matthew 24:34-This generation [**of souls**] will certainly not pass away [**die due to disbelief**] until all these things have happened. John 11:26- whoever lives by believing in me [then **their soul**] will never die.*

Hopefully that is somewhat clearer than mud. The main thing to recognize is that the Bible claims that self-centered pleasure will surely lead to misery. Quit it right now. Just

48

quit. Go out and help someone, truly learn empathetic love, and humble yourself and you will be happier, and experience a life beyond your dreams and your life force dimensional qualities will begin to help shape our reality. I know this part has been religious. That's all the preaching I'm doing, but please remember there are 3 distinct dimensions in everyone's life and they all are REQUIRED to build our reality. With this beginning and the religious flavor, let's go back and look at the various Dynamos in more detail.

Dynamos

I don't have any way of knowing if you are reading this and grunting or if I have extended your awareness a little, but the dimensional dynamo concept possibly is even harder to picture than the 1s dimensions. Wait just a minute! I did here you grunt just then.

The entire universe can be express by 4 distinct things- solids, forces, life, and time. Each is 3 dimensional and each dimension is mutually perpendicular so they cannot be directly perturbed by the others.

What if each of these components is not simply an effect but instead each is necessary to secure this universe? What if life was part of existence? What if electro-magnetics was needed to secure this world? The first thing you would probably question would be that if these things were "DIMENSIONAL" they would have to be able to be characterized in mutually perpendicular effect dynamos that prove their existence. How in the world can you attach 12 dimensions perpendicularly? The dynamo is the answer. Dynamo is just a word for a group of functional blocks working together so don't get wrapped up in the words. It's the mutually perpendicular thing that may be important here. In a vibrational system like a magnetic field, there are 3 perpendicular lines of force which are neutral to the other 2 forces. What we call energy should be looked at. To do this we will look at the energy equations and provide them with new definitions. Generally, these are the same equations you are used to but definitions have been adapted to the 12-dimensional universe reactions and care has been taken in not jeopardizing reason by eliminating many of the required elements of this universe because there must be order in physical laws.

Instead of the normal 3 dimensions with time as the activator, our world is really split into 4 sets of three dimensions. Each set of 4 works together similarly to how you had originally believed the 3 dimensions reacted, but separated along lines associated with vibration rather than spatial separation. As it turns out, the separation is an effect rather than a dimension. Let me give you an example. Length and width and height are all exactly the same dimensionally attributed components that are mutually perpendicular. While perpendicularity is a sign of dimensional separation, the reactions that cause perpendicularity must be reexamined. In a truer world matter is defined by spatial separation, gravity, and nuclear bonding. Length, width, and height cannot define all of these things, so a different definition must be considered. The same holds true when trying to understand light, life, and time. None of these can even remotely be defined spatially.

Mutual Perpendicularity

In this world everything that interacts but holds a definition unto its own must be mutually perpendicular to the other defining elements or they cannot be a true independent dimension. What I mean by this is that there are key elements that make up the universe. While they may have some similar characteristic, they also can act separately in the universe. It's that way with the length, width, and height elements and in the vibrational defined world the same thing holds true. The advantage we have is that vibrational elements are more dynamic that the original, extremely limited definitions of our universe. The diagram following characterizes the 3 dynamos and their individual perpendicular dimensions joined by the dimension of time. Of course this is not the way it actually looks. All of the dimensional strings would look, sort of, like vibrating circles as discussed previously, but it does describe the connective nature and the "free" dimensions called life-subconscious, Electro-magnetism, and Aether-Gravitation.

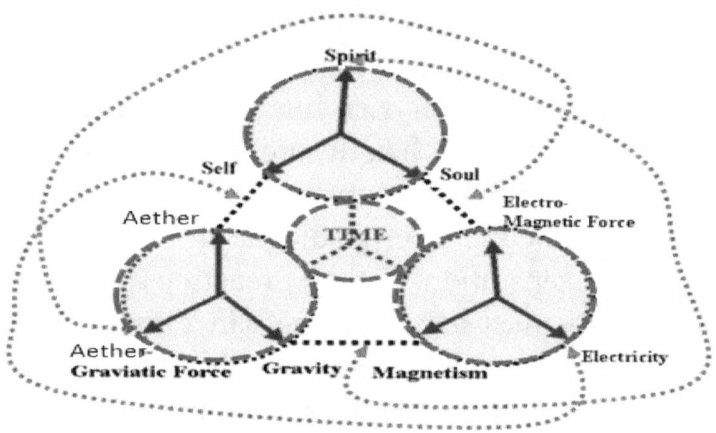

The dotted lines show how the open-dimensions tie back to the others to support similar characteristics. For instance, Aether, self and electricity are all potential dimensions that govern Matter, Life, and Photons. Each of these can be characterized almost identically with the other 2 dimensions in their separate dynamos. These dynamos are given "Life" by Space-Time. If "dimension" makes you uncomfortable you could also say Spirit Force, Photonic Force, Nuclear Force and the effects of time. I know you can't picture this as an object because spec-time disrupts what you normally think of as mutually interactive, but I assure you that space-time does exist and it changes to support what reality we are experiencing. Here is a hard part to understand, we are not all experiencing the same "space-time". A minute for one person may be different to another individual when they are not together. It may go fast for one and slower for another. As soon as the two are in the same space, everything is synced up. This is called red-shift when referencing matter and is called relativity when referencing life.

Why Three-dimensional Dynamos?

I think as we go along you will see why everything develops in 3 mutually perpendicular directions to produce reality. One example is very easy to see. We can and do see the electromagnetic dimensional dynamo and measure its effects. We know that without these effects we have no light and with no light there could be no true reality. By grouping the dynamos into 3 groups, each of the dynamos can also be mutually perpendicular with respect to 2 other dynamos as the 4th dynamo assures our universe does not implode on itself. By rights I should start discussions of dimensions by looking at electro-magnetics first, but I think starting with dimensions of matter might be best simply because we can see and feel matter.

Structural Dynamo

Everyone views particles or our "unified particle" that makes up everything as the epitome of the universe containing a volume, but by itself, it cannot construct a system and other things pop into mass that must be considered to construct a universe. The first thing we find is that all mass seems to have something we call gravity. Gravity is not usually defined as being part of a particle; in fact no one has really defined it well at all, because it confuses them. Even with the confusion, there it is. Now we know that both mass and gravity are needed to construct a universe; but wait!! How in the world do you make particles come together into clusters and hold them there? In fact, particles themselves have no real existence without gravity and a nuclear union. In fact, fermions really are doing the work associated with mass as a "Quasi-particle" that is missing the gravity portion. This is where the nuclear bonding comes in. It is like "particle paste" and certainly needed to build any mass even up to a mass we might see as a universe. The faster the sub-particles vibrate the more "PASTE" that is made which allows more mass to be defined. If gravity is not part of mass but requires mass, it must be based on a system that is <u>perpendicular</u> to the other components of mass. The effect of nuclear attraction is also characterized by mass, but it is not the combination of particles. It is something else. All three are required to establish a mass, but each must be reviewed separately as they have no real meaning to the others. Let me go one step further. Any change in the nuclear "force" or "gravity"

WILL change the characteristics of MATTER. In any dimensional dynamo each dimension can be recognized as a charge and its stability is typically established by mutual perpendicularity as shown below left. Because everything is based on a vibration, we must show these things as vibrational paths as shown to the right. I think I'll show them as straight lines and you can imagine the wiggles.

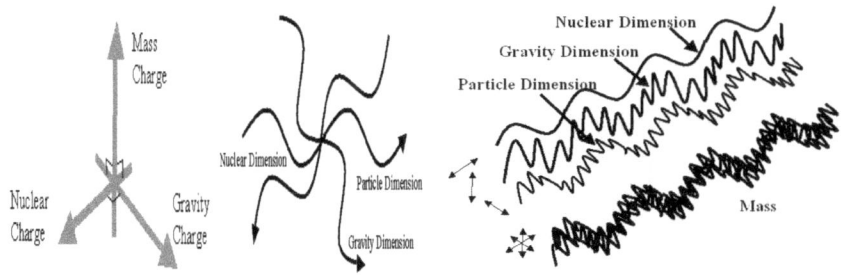

The third diagram shows the nature of the mutually perpendicular assessment on a linear plane. Each of the mutually perpendicular elements can work together to change characteristics. By all rights, I should have shown the vibrational paths as circles, but this may give you a feeling of the perpendicularity. If one views the string along only one plane [say the particle plane [more correctly the Aetheric plane, but particles are less confusing here], one would only see the vibrational aspect of the particle, not the vibration of the other 2 dimensions making the dynamo. Please see that without time there could be no vibration and matter would cease to exist. Also note that when space disappears, vibrational frequencies would go to infinity with similar results.

Entropy Explained

This section has equations in it. These are notional equations to show the connection between the various dimensions of our universe and to show similarities in dimensional duals associated with the universe. Please don't worry about them.

Entropy

A third thing will also be ventured. That is the definition of entropy. I know you have heard about the LAW OF ENTROPY, but <u>you have no idea where it came from</u>. I will tell you. Someone saw how particles separated as far from each other as possible and made it up. Well! In the 12-dimensional universe, there is a simple equation to confirm the law of entropy. In fact, because there are 3 dynamos working together in any specific space time that is experienced in this universe, there are three types of entropy that can be recognized. We give these things common names and see them in the lab and in life.

Energy, Conductance, & Resonance

If we wanted to define these dimensions mathematically, the equations would be of the dimensional energy, dimensional conductance, and dimensional resonance. These functions sound like electro-magnetic indicators because in a frequency-dominated universe, electro-magnetics become the operational indicators as will be cleared up in the details that follow [I hope!].

Dimensional Energy

You will see that all the various dimensional equations are defined by "energy".

> *Therefore, dimensional characteristic equations are of the same form as all universal energy equations.*

Please note the similarities as we go along to all energy equations that we attribute in our space-time defined world.

Dimensional Conductance

In the frequency domain, existence can be defined as the energy of its vibration. We also can look at the association of a dimension to vibration. In a vibrational sense we must understand how a dimension's frequency affects how it is represented in our universe. For that, we must characterize the dimensional element with respect to its dimensional "conductance".

> *The conductance formulas of dimensions are of the same format as all conductance formulae typically used in electro-magnetics as one would surmise because the electric energy and magnetic energy dimensions are two very important characteristics of our universe.*

Dimensions can be characterized either by the basic determination of Inductance [those that induce energy into a system] or capacitance [those that hold energy by relative position]. Dimensions that are capacitively distinguished affect dimensional conductance directly while those that induce energy affect conductance inversely.

Dimensional Resonance

In a vibrational world, vibrational patterns that are sympathetic increase in energy or sustainment. The opposite occurs at all frequencies not sympathetic to the base. We see this all around us. A vibrating crystal like the one that keeps your computer going, for instance, enjoys being at a set frequency of vibration and only a small amount of sustainment energy is required to keep the oscillations continuing. If a vibrational pattern is introduced that is misaligned or phase shifted from the vibrational pattern, the crystal vibrations will quickly diminish and die away. As mentioned before, this resonance is the basis for quantization of matter and here is the kicker. The dimensional energies act in the same way that the crystal does. It's not a mystery nor is it very confusing. I'm not going to address the various equations that are required to define this level of sustainment and resonance for each of the dimensions, but it is important to understand the concept.

A Dimensional Resonance equation is of the form
"$R = IC^{-1/2}$" where R is the associated resonance level and I and C are the respective levels of the Capacitive and Inductive components of a particular dimensional dynamo.
[I'll explain this in a little bit.]

Let me give you an example. In this case I will use the structural dynamo dimension component associated with a Fermion [which is the capacity of mater or contains the capacitive component] and Gravity [which induces stress on the Fermion or contains the inductive component]. As we go through this section I will explain this a little more. First let's develop our first set of dimensions and the equations that define them.

I know you have been told that it was something called nuclear force and pulling atoms apart causes these massive explosions and the atomic bomb was created, but that isn't the whole story. This nuclear force is simply the combination of Aetheric and gravitational forces that are adjoined at a resonance [comfortable place]. Change the characteristic vibrational pattern and the particle attains resonance <u>as a different thing</u>. Faster vibrations cause larger particles, slower vibrations cause smaller particles. It's all a matter of structural resonance. What we find is that the other two dynamos have this same resonance characteristic as well.

LENR Miracle

Something called low energy nuclear reaction is changing how we think of atoms. The basic concept of this new energy source is conversion of Nickel into copper or some similar combination by vibrating hydrogen ions with heat to the right frequency to pull energy away which forces a proton to be dropped out of a nucleus. As I described in the first book on Vibrational Matter, thousands are investigating the effect and designing ways to provide homes, factories, and vehicles with an almost endless supply of electricity for power. This simply vibrational change is drastically changing our life as this book is being written. One could say LENR forces a level of Entropy with a waste product of copper. In the old days, this would be called alchemy as the structural dynamo is affected most visibly.

59

Structural Equations

So as not to belabor this section, I'm going to provide you with structural "notional Equations" and you can build the others if you like. These are simple equation bases and all dimensional components have similar functional descriptions which can be characterized as similar defining equations. Luckily these constructs are not new to electronic engineers as the Electro-magnetic dynamo is manipulated often to make light, sound, and other useful forces and we can use similar equations due to dynamo similarity.

$E_g = \frac{1}{2} Gv^2$-The Potential Energy [E] of any dimension dynamo is the potential for expanding] I know you think this is the potential energy equation but in the vibrational world it works for gravity where "G" is the Unit of <u>induced</u> **Gravity** and "v" is the motion of gravity governed by the "distance from a center of mass".

$Zg = 2\pi fG$ -In the same form as "inductance" in the Electro-magnetic description this Z is always describing the ability of a dimension to induce an affect from universe [its inductance]. In this case we substituted the L [magnetic inductance] for a G to bring in the Induced Gravity. The larger the mass can be affected by the universe. Let me just share one more thing here. If the frequency of the vibrations go to zero, this dimension has NO effect on the universe.

$E_m = \frac{1}{2}Mz^2$ -The kinetic energy of any dimensional dynamo is of this form, where "z" is the controlling element of Mass [<u>Aetheric amplitude</u>] and "M" is the <u>mass capacity</u> of the system. If the frequency of vibration goes to zero, there is no energy associated with what we now call Aether.

$Z_m=(2\pi fM)^{-1}$-In the same form as "conductance" in the Electro-magnetic description this Z is always describing the ability of a dimension to affect the universe [its reactance to the universe], f is always the vibrational component and M is the Mass capacity. As the vibrational component of gravity increases, less and less gravitational inductance is required to affect the cosmos. Not only should it hold true, but also corollaries to it should hold true to similar dimensions in the other 2 dynamos. Strangely if the frequency of vibration goes to zero, this entire equation is not defined.

Please notice that the gravity and mass equations rely on each other to exist. We would find that all Induced and Conducted dimensions act this way.

Nucleatic Waves or Matter Waves

The third structural dimension is slightly different in that it is the transverse of both gravity and fermionic waves. This is what established the thing we call resonance as described next. When we put the energy equation together we find something very strange. The ½ goes away.

$R= GM^{-1/2}$-In the same form as "resonance" in the Electro-magnetic description this R is always describing the ability of a dimension to sustain its properties. G is again, the induced Gravity and M is the mass capacity. Please note that both must decrease to increase resonance. Particles want to dissipate. If it were not for something people typically call "nuclear attraction", this resonance equation tells us that all particles would diminish to their highest state of disorder. Particles would get smaller and smaller. Wow!

We just found the LAW OF ENTROPY and why entropy or disorder fits into the universe. I'll bet you never understood why entropy occurred, you just understood it had to be here because the "LAW" told you it must.

E_n=MC^2: This is the Mass resonance Equation [universal Law of Nuclear dimension]. It is derived from the following:

$$v=C/\lambda.\ E=B/C\lambda.\ and\ M=B/\lambda C^3$$

If "B" is a boson constant and v=frequency. Boson constant is an equilibrium that insures gravity and particles can be freely interpreted. Boson constant is an equilibrium that insures gravity and particles can be freely interpreted. This can also be known as the Nuclear Energy Unit]. The "C" is the speed of nuclear action [also speed of light].

Einstein's Missing 1/2

The "½" typically associated with this energy equation must be removed from the transfer dimensions of each dynamo as they actually live between both universes and share dimensional capacity from both. That's why Einstein was getting all his equations to work out. These energies work together as a unit in a general form so we think of them as one thing, but each can be addressed separately, in fact, we can affect one of the dimensions in a dynamo without affecting the others, but it is not easy. All dimensional dynamos have Induction, Capacity, Resonance, Potential energy, kinetic energy, and the oddball transfer energy. By simply changing vibrations, everything about a particle, electromagnetic wave, life and space time changes. While it may like a particular resonance level that "it" likes, it can be forced to a higher frequency level to allow it to affect our universe more, produce more force, and expand in space. The really nice thing about our universe is we can take the same equations and find out what will happen to the other dynamos. We will get into the increase of resonance of living people a little bit as people control our reality much more than they think and as long as they don't try, they cannot affect it.

Frequency and Energy

In all dimensional strings, we find that the higher the frequency, the more attractive the particle or reaction, the more dynamic the frequency [more amplitude] the more energy that can be established, but in the structural dynamo, a particle itself can be expressed as a **single dimensional vibration**. This is the mark of a dimensional string. All 12 dimensions can be expressed this way.

All dimensions are emanations of vibration and all are characterized as something we call ENERGY.

Quantum & More Quantum

While some indicate that dimensional strings can be either closed or open, I cannot get my head around the open strings. Therefore, in this study, only closed loop strings are considered because with the closed environment vibrational quantum can be established and everything in this universe seems to operate with specific quantum. That is that only certain vibration frequencies can be understood within a closed string given a particular time base constant. Not only does it make it easier to understand the regenerative nature of a vibrational string, it also helps define the reasons that there are not an infinite number of types of atoms, or an infinite number of possible consequences from a reaction, or why light is quantized into sections associated with how the electromagnetic reaction affects the other elements of this universe.

63

Think of it like a track that is a certain length and a wheel riding on the track represents a vibration with a single point on the edge of the wheel as the vibrational realization. As the wheel moves down the track, the motion of the point on the wheel describes a sine-wave characterized by the wheel size. In this quantum restriction, the point on the wheel must begin in contact with the track and end in contact with the track, as shown on the following diagram left.

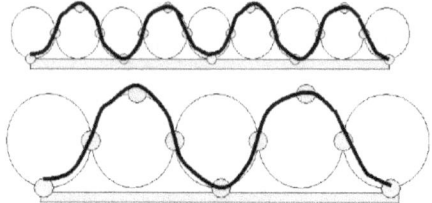 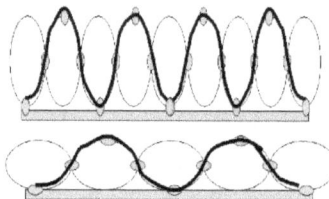

Notice that both sets of wheels cause the "point" to be on the track at both ends of the track. The ends of the tracks represent the closure of a string [Put the 2 ends together to make a circle.]. The number of times the "point hits the track represents the quanta of string. The length of the darkened string in both diagrams represents the energy of the vibration. In these two examples the energy level is about the same, but in this world, vibration can be defined by more than just a simple wheel, as shown to the right. As energy is put into a string, the vibrational length gets longer as shown on the first diagram below and as energy is taken away the vibrational length get shorter when compared to the at rest patterns shown above. From these diagrams we can determine some axioms.

Given constant vibrational amplitude in a close string, the energy level goes up as the frequency goes up.

Given a closed string, changing the vibrational frequency to a higher level requires input of energy and a release of energy occurs when the frequency is decreased.

All matter is defined by characteristic number of quanta needed to fill in the closed dimensional string.

Please do not worry about the equations. As I said they are only notional as many of these parameters are impossible to measure. I only wanted to show these constructs to demonstrate that all dynamos operate the same. I will not go over the others except as a table to show similarity.

Defined Quanta

Many different vibrational elements can reside on a closed string, but all must have a characteristic quantum of the string.

As shown above, the string is made up of 2 different vibrations with a quantum of 4 [4 cycles of the vibration are completed along the "string length"] and one vibration with a quantum of 2. The lower the quantum the less effect a vibration has on the system. In real life the quantum numbers are enormous for the more powerful elements. The heavier elements vibrate at a much faster rate so they become more powerful, with simple systems vibrate much slower and are more insignificant. We see the same effect with light. As the frequency increases, the electromagnetic wave increases in power until the cosmic rays of the highest frequency levels can rip things apart.

Let's just see what the energy equation does for us. As the Energy in the system goes up, 2 things can happen, either the vibrational frequency increases or the mass of the object get denser. This is accomplished by attracting more fermions together to form quantized "atoms". Each of the atoms would have a specific vibrating frequency to Mass ration. In an entire system such as a universe, many fermionic dimensional strings make up all particles that can be seen.

Mass Invisibility

As discussed in the vibrational matter book, matter in this string can be visible or invisible. If 2 fermionic strings with identical energy level and frequency come in contact [out of phase as shown on the following page], the result is that there is no effective energy in this dimension associated with the strings. They cancel each other out, so to speak. They still exist, but there is no outward sign of that existence. Simply change the phase of one and both become visible. At an event horizon of a black hole, this has been described. Also in spallation units, gravitons have been produced which typifies invisible mass.

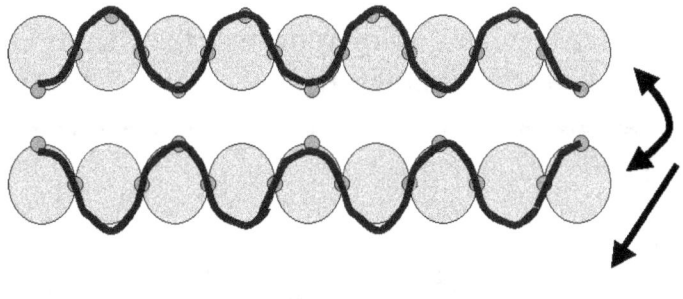

Wow! That was easy.

Attraction

Nuclear attraction is vibrational and like the fermionic dimension, out of phase interactions appear as invisible nuclear forces, but in-phase interactions cause strong attractive forces. If a proton vibrating at a different, but close frequency to the original set of bosons, the resulting vibration will be changed. This change will result in some of the bosons having to leave to insure equilibrium and a new mass will be defined.

Spallation

The atom smashers of today are really neat, but what are they doing? By adding 2 different frequency-fermionic-strings together, the frequencies should beat together. One would assume that there would be 2 beat "fermionic" frequencies and the 2 original frequencies. In a spallation chamber protons are smashed into isotopes to see if the atomic mass can be changed and, sure enough, one of the beat frequencies shows up. The chart following shows the typical number of different materials that have been made by smashing atoms. Using heavier particles in the first place makes it easier to produce heavier atoms and the mass is always less than the original mass. The closer the material is to the needed material the more atoms produced. For instance, gold has an atomic mass number of 196 so you can see it is easier to make gold from lead than from Uranium, but both can be turned into gold.

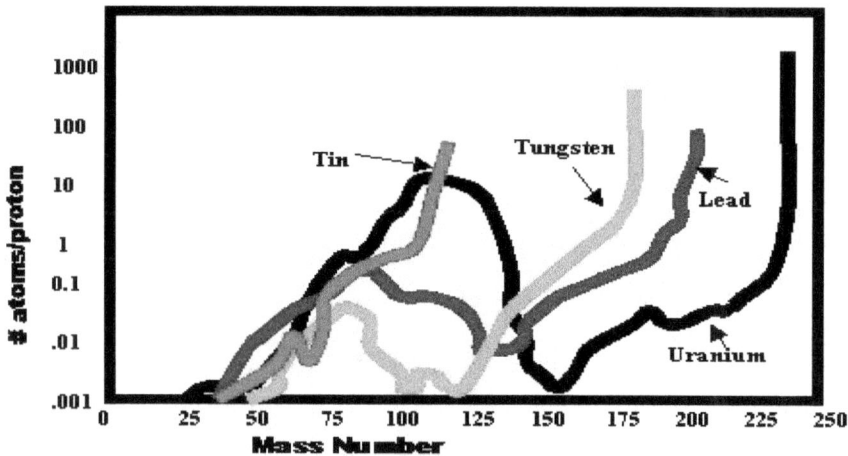

From the above we can assume that it would be easier to make gold out of mercury or thallium than lead, but let's see what is happening here. The accelerated proton hits the bosonic mass and changes its characteristic vibrational

68

pattern; some of the bosons must leave to insure equilibrium and gold or just about anything is made. There is another curious thing to notice about the preceding graph. There is a secondary bump in atoms created with atoms having about 50% to atomic number of the primary structure. While this probably has something to do with relational quanta of what I call the Aetheric dimension, there has not been an explanation that sufficiently describes this unusual trend. For now it is simply interesting so we will go on.

Particles

While these three dimensions are separated from the next 6, they are related. I left the 3 dimensions of time off on purpose. Remember the Structural dimensions are mutually perpendicular nothings vibrating away from some central node. Billions of these nodes make up the central cores of all things we see as atoms. The question might be where do the "nodes" come from? We will get to that as we investigate the Operational dimensions.

Before we get to the next I want to show you a table of dimensional equations to show how they all fit together and characteristics that show simple actions within each of the dimensional dynamos. Therefore I will not be addressing the equations in the other dimensional discussion. As you would expect, the table shows consistency with known energy, reactance, conductance, capacity, induction and relative mass equations in use every day. The dimensions have these similarities to allow them to work together to form our universe.

Structural	Operational	Life	Time
$E_m= \frac{1}{2} Mz^2$	$E_e= \frac{1}{2} Cv^2$	$E_l= \frac{1}{2} Ay^2$	$E_t= \frac{1}{2} Rx^2$
Em=Mass Potential	Ee= Electrical Potential	El=Life Potential	Et=Time Potential
M=Mass capacity	C- Electrical capacity	A=Anthropic Capacity	R= Time capacity
z= Aetheric amplitude	V= Electrical Amplitude	y= Life amplitude	x= Space distance
$Z_m=(2\pi fM)^{-1}$	$Ze=(2\pi fC)^{-1}$	$Z_l=(2\pi fA)^{-1}$	$Z_t=(2\pi fR)^{-1}$
Z= Aetheric Reactance	Z=Electrical Reactance	Z=Life Reactance	Zt=Time Reactance
$E_g=\frac{1}{2} Gv^2$	$E_m=\frac{1}{2} LI^2$	$E_s=\frac{1}{2} SD^2$	$E_h=\frac{1}{2} WG^2$
E=Kinetic Gravity	E=Kinetic Magnetism	E=Kinetic Soul	E=Heaven Energy
G=Induced Gravity	L=Magnetic Induction	S= Soul Induction	W= Heaven Induction
v= Gravity distance	I= Magnetic Control	D= Soul Control	G= Heaven Control
$Z_g=2\pi fG$	$Z_m=2\pi fL$	$Z_s=2\pi fS$	$Z_h=2\pi fM$
Zg=Gravity Reactance	Zm=Magnetic Reactance	Zs=Soul Reactance	Zh=Heaven Reactance
$E_n=MC^2$	$E_p=CC^2$	$E_f=AC^2$	$E_t=AC^3$
E=Mass Energy	Ep=Photonic Energy	Elf=Life-force Energy	Est=Space-time Energy
$R= GM^{-1/2}$	$R= LC^{-1/2}$	$R= AD^{-1/2}$	$R= AR^{-1/2}$
R=Mass Resonance	R=Photonic Resonance	R=Life Resonance	R=Time Resonance

As this is time consuming and cause for confusion, I will not be building these equations for the other dimensional dynamos.

Operational Dynamo

Having matter is one thing, but the universe can only be placed with the previous dimensions. Matter cannot be used and it does not have defined. For that effect we must key in on the interactions of light, electricity, and magnetism. The union of these things has been known for many years. What many do not recognize is the triple relational grouping with photonic action being NEEDED to perform the other 2. While all are connected, the reason that the union is not noticed is that each affects the other in a perpendicular way [with respect to the vibrational mode set up to define the other elements of the operational factor. The diagram would look exactly like the one I drew for the Structural Dynamo, but the effect on the universe is substantially different. I will explain the characteristics of this dimensional dynamo by using equations that are almost exactly the same as those used to define matter. It is this similarity that helps us define the dimensional dynamos as a needed tool in understanding our universe. The three dimensions associated here are described below.

Magnetic "Dimension" appears to be a perpendicular vibrational component of Electricity. To put this in perspective, it has been determined that a magnetic field created by an electric current decreases by the square of the distance from the source. Also it should be noted that electricity is "CREATED" whenever a magnetic field is vibrated such that the field produced is continuously changing. Magnetism is said to induce force just like Gravity was the inductive part of mass.

Electrical "Dimension" is known to be perpendicular in nature to a magnetic field. Any electrical field will produce a perpendicular magnetic field. Like the magnetic duality, vibrating electricity CREATES magnetism. You might think that we could simply use one for creating the universe and simply vibrate it to get the magnetism thing. It doesn't work like that. Electricity is the capacity to make E-M or Photonic forces just like Aetheric dimension is the capacity to build mass.

Photonic "Dimension" or photonic emission appears to be the mutually perpendicular vibrational component of the Magnetic-Electrical [electromagnetic] duality. Like the third component of the Structural Dynamo, the photonic dimension is like "operational paste". We know that Photonic intensity created by electromagnetic dual decreases by the square of the distance from the source. This helps establish the relationship of the three Operational vibrational components of the dynamo.

Please do not tell me that the universe could still exist without photons and don't tell me that when a photon is a particle it is part of the first dynamo and when it is a wave it doesn't matter. Photons matter. Think about this. A photon is sometimes a wave and sometime a particle. Doesn't that should like its vibrating between one dynamo and another to stay in existence? We may, very well find out that nuclear "force" sometime affects particles and sometimes affects electro-magnetics or the life dynamo and we will begin to understand the universe better [What we will find in the third dynamo is that one of its Dimensions acts just like the Photonic dimension to the electrical and magnetic components.] The characteristic energy of photons have a similar $E=MC^2$ function as it travels the speed of light and goes between universes we can substitute the M for any suitable letter denoting the Electrical capacity. Curiously, while magnetism doesn't seem to have a "basic unit" [theorized as the monopole just like the Gravity thing above doesn't have a basic unit that has been found. Later we will see that life also has no recognizable "unit" as well.]

Let's see what this means in the world of vibrational matter. We know that a changing magnetic field creates an electrical field and a changing electrical field creates a magnetic field inversely. We know that magnetic field strength decreases by the square of the distance to the host electric field. We also know that "changing light" changes the electro-magnetics of the system around the light inversely. We also know photonic intensity decreases by the square of the distance to its electromagnetic source. All three components act as one "dynamo". A vibrational change of one affects the others in some inverse way to keep the equalization of the "operational dynamo". The graphic to the left below shows the relationship of the three partners. The diagram on the right shows how they are characterized in the vibrational world. The vibrational frequency of each governs the characteristics of the entire group.

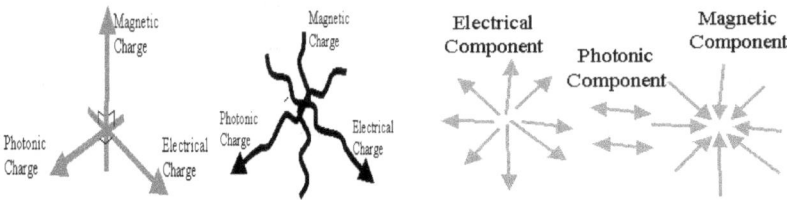

It should be noted that this system or Dynamo does not create the particles associated with mass. They create the operational functions that are needed to perform work, but in a time space model, they have similar mutual exclusions with respect to interactions.

73

Electro-Magnetic Resonance

While I'm not going through the similar equations again, I do want to describe the Resonance dual. As far as resonance, the equation is simple and previously described. I'm simply going to write it here and describe how it will hold true in the other dimensions. $R=(CL)^{-1/2}$--- The highest level of sustainment happens when the electrical capacity and magnetic inductance is at its lowest level. Think of it as electromagnetic entropy. It is where electromagnetic waves feel most comfortable in this universe.

Photons

We already know by experimentation that the higher the frequency of vibration the more powerful the photon becomes and this describes it easily. If we modify the frequency of the photon, the actual photonic charge will be reduced. Form equation dualities we see that if one can decrease the photonic storage capacity, the photonic energy also will go up. Aha! We seem to have 6 dimensions in our universe and time-space that we will talk about later. Besides these things, there is something missing that people ignored for a long time and simply described many things as anomalies. That something is **life** itself. I know you are generally accepting the 6 dimensions and space- time pretty well, but the life thing is way too out there so I had better provide an entire chapter on this important dynamo. Before we can get to it, we need to look at Black Holes.

Black Holes

As I mentioned before, an object that is all gravity becomes what we call a black hole [window to our linked universe]. We can sense that as similar Magnetism-only thing [monopole] sort of exists as well and it too could be considered a "Window to our linked universe". What we will find is that the 2 are identical and opposite. In an adjacent universe, an opening "from" a black hole becomes one of these Magnetic monopoles while a Black hole from our universe becomes one of the Magnetic monopole things. To understand this association, one must recognize what an "in-wave" is, according to Dr. Wolff and others. It seems that all out-waves from an adjacent universe bombard our universe as "in-waves" through these "windows". Also we should note that In-waves are completely out of phase with those escaping our universe. Any place there is a "collision of vibrational peaks", the vibration is nulled and that point begins to build out-waves. As mentioned before, this represents the core points of all atomic structures. Particles in our universe become electromagnetic forces in adjacent universes and vice versa. Let me just state this plainly.

We cannot have matter here without having an adjacent universe.

This can be determined by a number of reasons.

- No "null points" could be provided which would start an out-wave generation.
- Because no energy, matter or life can be created or destroyed, there must be a regenerative process where in-

75

waves from another universe can be allowed to regenerate energy leaving the universe.

- The apparent regeneration of matter at the "event horizon of a black hole makes no sense.
- The Theory of super-symmetry which allows for major reorganization of energy by having an opposite effect in an adjacent universe makes no sense.

I briefly mentioned this as well, but not only does this universe have to be attached, it also must have its time backwards to us.

Our linked universe experiences time backwards.

This will make more sense as we discuss the Time dimensions, but let me first give a brief introduction to the Life dimensions.

Life/Death Dimensions

Let me talk briefly about the three dimensions of life that turns a lump of DNA that has been energized by the electromagnetic forces of our perceived reality into life. We will look at it some more later as it is a very odd set of dimensions so I will spend more time with them to help that part sink in after we study time. Life requires the combination of self, soul, and spirit as described in the Bible. Don't go thinking that because this book has excerpts from the Bible that it is not a science book. There are other examples that we could use, but many already know the Biblical history.

*1 Thessalonians 5:23- "And the very God of peace sanctify you wholly; and I pray God your whole **spirit and soul and body** be preserved blameless unto the coming of our Lord Jesus Christ."*

Over and over and over again the Bible tries its best to introduce everyone to the three dimensions of cognitive life of a person. Here are just a few. When you are being told these many times, we are made of three separate entities, we should listen.

*1 Corinthians 12:13 -For by one **Spirit** are we all baptized into one **body**, and drink into one Spirit.*

*Matthew 10:28 -And fear not them which kill the **body**, but are not able to kill the **soul**:*

*1 Corinthians 6:20 - glorify God in your **body**, and in your **spirit**, which are God's.*

Ephesians 4:4 -*There is one **body**, and one **Spirit**, even as ye are called in one hope of your calling;*

Romans 8:10 -*And if Christ be in you, the **body** is dead because of sin; but the **Spirit** is life*

Romans 8:23 - *they, have the firstfruits of the **Spirit**, waiting for the adoption, to wit, the redemption of our **body**.*

Daniel 7:15 -*I Daniel was grieved in my **spirit** in the midst of **my body**, and the visions of my head troubled me.*

Isaiah 10:18 -*And shall consume the glory of his forest, and of his fruitful field, both **soul** and **body**:*

Isaiah 51:23 - *they which have said to thy **soul**, Bow down, and thou hast laid thy **body** as the ground,*

1 Corinthians 5:3 -*For **I [soul]** verily, as absent in **body,** but present in **spirit**, have judged already,*

Romans 8:13-*For if **ye [soul]** live after the flesh, ye shall die: but if ye through the **Spirit** do mortify the deeds of the **body**, ye shall live.*

Micah 6:7 - *shall I give my firstborn for my transgression, the fruit of my **body** for the **sin of my soul**?*

James 2:26 -*For as the **body** without the **spirit** is dead, so faith without works is dead also.*

Job 7:11 -*Therefore **I [my body]** will not refrain my mouth; I will speak in the anguish of my **spirit**; I will complain in the bitterness of my **soul**.*

Numbers 19:13 -*Whosoever toucheth the dead **body** of any man that is dead; and that **soul** shall be cut off from Israel:*

Matthew 12:18 - *my **soul** is well pleased: I will put my **spirit** upon him,*

78

*1 Corinthians 15:44 -It is sown a **natural body**; it is raised a **spiritual** body.*

*1 Corinthians 7:34 - she may be holy both in **body** and in **spirit**:*

*Hebrews 4:12-For the word of God is a sword, dividing asunder **of soul** and **spirit,** and of **the joints and marrow**, and is a discerner of the thoughts and intents of the heart.*

*1 Samuel 1:15 - I am a woman of a sorrowful **spirit**: I have poured out **my soul** before the LORD.*

*Isaiah 26:9 -With my **soul** have I desired thee in the night; yea, with my **spirit** within **me [Body]** will I seek thee early:*

Let's define a Cognizant Life

You probably noticed that all dynamos had something in common. There was a "potential" for its essence when there was no vibration and when it vibrated too fast, it sort of went "beyond our reality" as pure gravity or pure magnetism or the spirit of a living person. The spirit is that part of life that goes beyond our reality. If you have heard about heaven, this is sort of a tunneling element for that place. The "Self" is the potential for life and the "soul" is what can goes beyond our reality as we venture away for this thing we can call self. Let me stop here and look at a very difficult statement presented in our Bible.

*Whoever wants to save his SOUL will lose it, but whoever loses his SOUL for me will find it" **(Mathew 16:25).***

Remember that little girl that picked up the car off her dad? This is sort of saying the same thing. Our soul is, *generally,* tied to this reality. This reality is kind of like a governor on a motor that keeps it from going too fast and destroying itself. It builds a reality around something that has been called the

79

"resonance of life". I'll get in this "resonance stuff" a little later but, essentially, everything in the universe is comfortable at a certain Main vibrational level. Remember, if matter vibrates too fast, it ceases to be matter and the same is true of energy that becomes magnetism. Well--Life is the same. If we [or our self-soul wave] vibrate too quickly, we will cease to be. That is, we will cease to be in this universe. The faster we vibrate, the more we control our surrounding "reality" just like mass gets more powerful as it vibrates faster and electromagnetic waves become more violent forces like gamma waves, etc. Everything in our universe is tightly affixed to everything else, but that does not mean everything is RIGIDLY defined.

If we can somehow vibrate faster than the "life reality around us, we have more control over the reality, but at the same time less interest in this reality so it is a two edged sword. On a small scale, you would recognize that if someone is "meditating" he loses his awareness of the things around him. That's what Jesus was stating in the book of Matthew. The person who, for an instant, gains tremendous power to help someone out from under a car is done much the same way.

Walked on Water and Move Mountains

Don't believe Peter, and Elisha, and Elijah, and Jesus didn't walk on water because someone said it's impossible. Don't think that you can't leave your body simply because it seems odd. Don't think that when Jesus told his people that *faith as small as a grain of mustard-seed is all that is required to move mountain*s was a lie because you had no understanding of what the word "faith" was as you kept getting it confused with religion. Certainly there were descriptions of "Faith-in God", "Faith in Jesus", etc. but this verse was talking about

80

"normal" faith or it would have been said differently. Let me just tell you the Biblical description of what Jesus told his followers. This comes from "Hebrews".

Hebrews 11:1- *"Faith is the <u>assurance of things hoped for,</u> the <u>evidence of things not seen</u>"*

This positive feeling increases the vibrational level of all three dimensional elements of our being. Jesus told his followers that "NORMAL Faith" could change the construct of reality. That was not faith in Jesus, he was telling them about Anthropic Science before anyone had even heard of it. He was God incarnate, but he told us *"We" can move mountains just by telling them to move if he says we can.* In the Anthropic expansion of physics, it is not only possible, it can be considered a requirement that is stifled by people simply "enjoying" carnal life with its sex, satisfaction, pride, power, and all the things that we begin to lose as we gain power to control our environment. One must abandon, self, sex, and survival to allow ANY modification of the vibrational level of life, so to speak, and allow a better control over one's reality including the reality of what happens after what we call death.

Only Meditate With Direction

Before we talk about death, let's briefly discuss something ominous. The verse indicates that *"only those who lose their souls <u>FOR GOD</u> will find it"*. By definition if nothing else, life must be indelibly tied to the Creator God [Whatever you want to call him]. I know I stepped around death for a minute, but let's go on to something we can describe as the Time Dimensional Dynamo.

Time Dimensions

We will look at our "Normal" forward time, our linked universe times which is mutually perpendicular, but backward in the absolute and something I call Lateral time. Like all the other components of our universe if we lose any one of these dimensions, our universe will not exist. I think it is easy for everyone understand the concept of forward time being required for reality, but the backwards time must also be dimensional as there is a big problem with non-replenishing forward time traveling to the ends of our universe, never to be seen again. We must have a regulator that pushes into our universe time just like it pushes in matter that we eventually lose. Like ALL dimensions there is a transverse dimension that cuts across both to allow their union and to allow for the travelling to our linked universe [just like a black hole allows for a matter transfer]. This dimension is called lateral time in this book and it allows for something called time travel [outside of time-space].

Conservation of Time

Like it or not, time cannot go just one direction. If time was always "advancing" soon all-time would leave the universe and soon there would be NONE. I know that sounds silly, but there is a simple method used in our universe to all for conservation of time, just like everything else. That secret is an adjoining time-reversed universe. While some call this universe heaven, it makes no difference to me what you call it. The three-dimension particle dynamo [this is where mass and gravity are generated] causes vibrational flow through the universe according to Einstein and many others. This set of waves is called out waves in that the soon reach the end of the universe and can be lost forever if it wasn't for another dynamo of dimensions we can call the force dynamo [this is where electro-magnetics is generated]. As the vibrations from these [in-waves] coming into our universe to build stresses are generally exactly like the out-waves except they are backwards in time. Guess what as our particle [out-waves] leave out universe they become "stress in "Heaven" and produce force and the Particles from outside our universe become those in-waves we need so desperately to hold masses together. Let's not look at life for a minute and just try to observe things happening in our adjacent universe. Everything would be going backwards.

Things would be getting younger and newer to us as time over there is backwards.

As time leaves our universe, it is sensed as backward time in our neighbor and vice-versa. We always have time and they always have time. String scientists call all this stuff "super-symmetry". I'm not just making stuff up---you know! The diagram following, hopefully, will describe what a more up to date description that these string theorists have put together and seem to work. The universe on the left might be ours while the one on the right might be the one we call heaven. While there are billions of mass vibration centers, I only drew one at the boundary of our universe. As it travels into our neighbor, the same thing [according to super-symmetry] is coming back at us, but everything is backwards so it forms stresses in our universe that we call Electro-magnetic forces that hold everything together and allow particles to build.

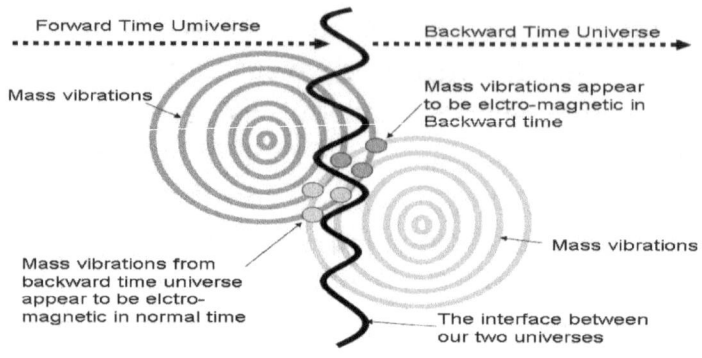

Forward Time Umiverse

Backward Time Universe

Mass vibrations

Mass vibrations appear to be elctro-magnetic in Backward time

Mass vibrations from backward time universe appear to be elctro-magnetic in normal time

Mass vibrations

The interface between our two universes

Each one of the dots drawn where the two vibration rings intersect can be considered a particle seed. From these, additional vibration rings will emerge which represent the combination of the out-waves and the in-waves.

All mass is made this way.

By the way, things would not feel backwards to people living in the adjacent universe if you were worried about them. To them, our life starts at its end and has us getting younger and younger. I know you were confused when Einstein told us that if you go near the speed of light, you quit aging and some of you understood that going faster than that would mean you would go backwards. This is the same concept, but scientists differentiate backwards time as an undefined reaction as density goes to infinity along the reduced speed timeline.

One way to look at it would be that particles become electromagnetic waves above the speed of light.

With that, let me bring in Dr. Milo Wolff. Dr. Milo Wolff, sort of, started where Einstein left off in defining what light, matter, and consciousness really were. While Einstein had initiated the thought that matter, light and consciousness were simply undulating nothingness [made of Aether], Dr. Wolff added a new level of insight that will get us a little

84

closer to understanding what all this is and then we can understand Anthropics which will ultimately allow us to release our consciousness. He gives us a defined picture of matter and light that is completely vibrationally based and uses the in-waves [entering our universe] and out-waves [Leaving our universe] as Einstein had said.

Explaining the Perception of Matter

Wolff-It is then quite simple [I hate it when someone says that] *to show that: The discrete 'particle' effect of matter is caused by the* **Wave-Center of the Spherical Standing Waves.** *The discrete 'particle' effect of light is caused by discrete Standing Wave Interactions/ Resonant Coupling.*

Dr. Wolff has completely separated the physical characteristics of the universe into NODES or intersections or what he calls "Wave Centers" of these wave things Einstein tried to characterize as Aether [Undulating nothings]. By describing them as spherical, we can see that while they travel at the speed of light, they really don't go anywhere.

Just to be clear let me tell you that we are talking about Aether, the "potential" for matter, vibrating. Only vibration establishes existence; once something stops, it ceases to exist. This includes Matter, light, life, and time.

Wolff Defines Time

Wolff-Time is caused by Wave Motion, as spherical wave motions of Space which cause matter's activity and the phenomena of time.

While this sounds like gibberish, it does something for us. It eliminates physical attributes. Now that the "length, height and depth" things do not stifle us, we may be able to define

85

light, life, matter and even time. Right now just sense mass as these ever-growing spheres of **undulating nothingness** [as Einstein would have said]. Here is Dr. Wolff again.

Quantum Entanglement

Dr. Wolff uses his same omni-observed definitions to show how everything seems to go towards specific quanta.

Wolff-Quantum Entanglement is likewise caused by the Interaction between the In and Out-Waves and all the other matter in the universe, thus matter is always subtly connected to other matter in the universe (i.e. matter is large not small, we only see the Wave-Center and have been deceived by its 'particle' effect).

If you remember I used the term Quantum fluctuation for all vibrational potentials for matter, life, and energy while Quantum entanglement is the interaction between quantum fluctuations. Sorry for the craziness, but this is really a new language.

Consciousness Controls Reality

Here is the important part to remember. All matter, light, time, and consciousness are subtlety connected. Changes in the mass we can call the common consciousness will absolutely change matter, light, time, and life. Consciousness can control REALITY. Before you go crazy, let me introduce a chart. This generally shows how Anthropics [Science of cognition controlling reality] works and as such how releasing our consciousness [either after we die or while we are alive which some call faith] can affect our universe.

- Please note that as matter [out-wave] leaves our universe, it becomes Energy in-waves and the opposite occurs for matter that leaves our linked universe [Heaven].

86

- Time goes forward for us in our universe but it appears to backward to us in the linked universe.

- Time has no effect on souls [Released consciousnesses] or spirits [Interlinks between universes].

- It appears that souls and spirits are locked in their respective universes, but, both can "travel" to the other <u>as an exchange</u>. A soul can travel to Heaven and become a spirit and vice versa.

A clearer graphic may be something like that below.

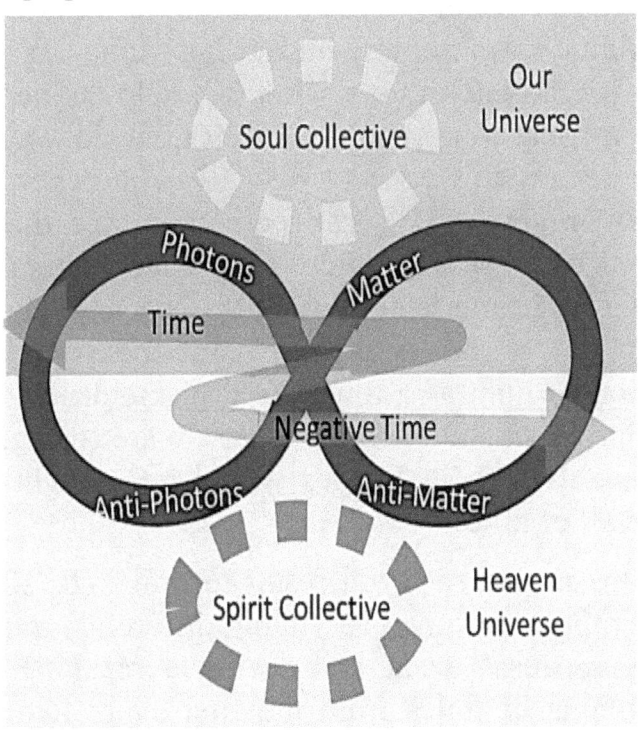

Lateral Time Dimension

Aristotle, Einstein, and Milo Wolff have tried to establish just what matter, light, time, and life were and while doing it they began defining Anthropics. The first step to understanding Anthropics and what time is may be something I call lateral time. This will help define light that is sort of misplaced in time, let's shift time sideways. As we do it, we review this concept we will get a better image of what light might be. We can potentially see the changes established by light if we view what I term lateral time. A person viewing time laterally would see the <u>beginning and end of time "at the same time"</u> and he would see the <u>progression of light as a time based phenomenon</u>. It would be difficult for us to determine what we are seeing, because we are not used to it. Images would be of light and you could see all of time.

I know this is odd so I will result to drawings. Remember this is simply viewing everything differently rather than initiating a causal event and seeing a resulting opposite event millions of miles from the first event like quantum mechanics has presented, this should be easy. I put a lateral time diagram below. You can see that God can see the beginning and end of your life simultaneously and everyone is piled on top of one another spatially. The excursions of the graph <u>represent light going forward and backward in lateral time's "equivalent of time"</u>. We could call this equivalence "mass resonance change" in our normal time perspective. I've labeled the viewer as God, because he may be the only one that can perceive this thing. From this vantage point, everything that happens to you from the birth to death are all shown up in one instant. There is no future or past, there simply is.

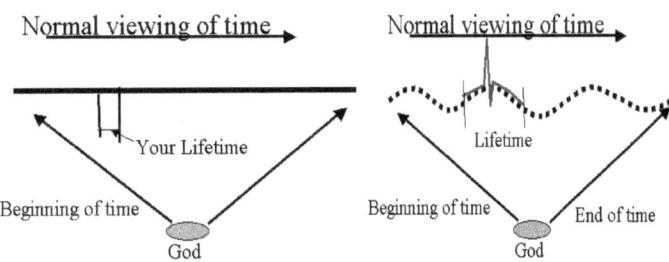

Time may be vibrational like all the rest of the dimensional strings of this universe. Instead of a straight line, on the to the right, I am showing it vibrating just like everything else does. As we go through time, the hills and valleys don't mean anything to us, but the variations could be witnessed laterally. The hills and valleys might be certain cyclic pressures like destruction periods, Ice Ages, wars, and other things that mark the cyclic nature of time and God could look at all these peaks simultaneously.

When God said he knew the beginning and the end, He would have had this lateral view of everything. He would see, for instance, your life all at once and in place of time, he would see modifications of resonance represented by the curvy line across the center of the viewing. Here is the part that is interesting. According to time dilation, anyone traveling or vibrating at the speed of light would sense this lateral view of time as all time stops for the traveler. I know this is all new to you so I will try to explain it as much as I can so you will appreciate what light really is.

Now for a hard question; *if everything is really made up of vibrations only, as I have presented, what would everything look like when viewed in **lateral time**?* Hurry up; the clock is ticking. -----Come on!

*If vibrations are emanations of modification over time, the answer would have to be a **"solid mass"**?*

This solid mass isn't a mass at all. It is simply a compressed vibration. OK! I don't know what compressed vibration of nothing really is. It is easy to write about it and make you think I know something special, but it is quite another to be able to picture what light is in your head. It all has to do with perception. As we look at lateral time in more detail we will see a stronger relationship between life and light.

*One can say that life is light viewed sideways and **light is life viewed sideways**.*

Certainly, no one would suggest that a light bulb had life, but what one may find out is that there are a number of similarities between light and life and some critical differences. Before I get back into these more exotic descriptions again, let me back up a little and try to reintroduce light as defined in "normal-time". Then we will combine it more in the lateral time world and finally, I will add in the more difficult theme of "Life Definition".

Basically, we can either look at time as a constant, as this impenetrable limit, or as something else. In this chapter I am going to bring out more of the "something" else. We can easily imagine time going forward with the past being behind us and, potentially, we can understand the concept of backward time, but what I need to expand here is sideways or lateral time. This is not to mess with you, but because

90

several areas in the discussion of light requires a limited understanding of this lateral time.

Vibrating Time

Before you can really understand time, you must first see it as one of the dimensions that make up what we call a universe. As such, it must be made up of the thing that makes up all things.

NO!

I don't mean atoms or bosons or even the almost unperceivable fermions. Time is not made of particles.

Hopefully you already guessed it. What I really mean is vibration.

We accept the vibration of light and the other things around us, but we typically view time as this straight line thing starting at the time we are born and ending when we die. We can sort of extend this same "time-line" from the beginning of the BIG BANG thing until the end of all time, but it still is in the same direction and has the same constant linear dimension. Oh!! How comfortable this description is. Very visible and very easily described, but then Photons could not go backwards in time.

Vibrating Electricity and Light

If electromagnetic fields didn't vibrate, they simply would not exist as would light itself. What I mean is that if you ever stopped a photon, you would be holding nothing. The bad part of this stopping vibration is that if we stopped vibrating, we would no longer exist as well. If we look at atoms, current studies indicate that they are simply clumps of common vibrational nodes rather than true substance.

91

Light Seen Laterally

While lifetimes take up a section of the time line, typically, light goes the speed of light. It is here one instant and gone the next only to be regenerated and be found again for another instant. Light doesn't actually travel along the time-line. As Einstein predicted at the speed of light there is no time. Light would view the universe LATERALLY as shown in the next graphic. Light would appear the same as life appears when viewing time normally. Normal mass would have no apparent mass equivalent and light would be represented as having this equivalent of life.

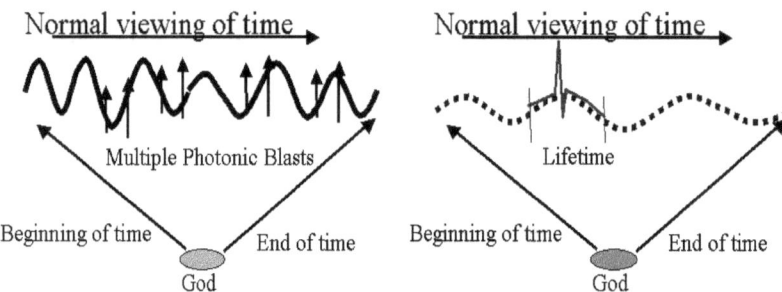

Speed of Light Example

That was the easy description with no meat. Let's put on some meat and see what happens. If a person leaves here in a rocket going the speed of light and returns going the speed of light, what would the rocket look like? The answer is that the rocket would gain infinite mass along the direction of travel. It would look like a beam of light and it would not travel on the "normal timeline. The rocket and the person experience LATERAL TIME as shown above right.

The time-line was expanded a bit to show detail the spike in the middle represents what happens when the rocket goes close to the speed of light. If a person could see what was happening outside, he would see everyone's life passing in an instant. The downward portion of the spike is his return to normal home at close to the speed of light. If you haven't seen it from the earlier examples, let me tell you that you will turn into--- "light"---- if you go the speed of light.

Turning Into Light

Let's look deeper. If you could see someone who was viewing you in lateral time, how old would they get as you aged? Of course they would not age a day because they could see your entire life as an instant. If you go the speed of light, how old do you get with respect to those not going with you? The answer is that you would not age.

A more defining answer is that if you could possibly go the speed of light you become light and are traveling in lateral time. However, all the particles in your body are vibrating so the particles making up your body are going backwards in time or they must stop vibrating.

Forget I said the backward time thing and let's move on. If particles stop vibrating, they do not exist. While this would be bad for 'LIVING" people. Light may have no reason to exist or not exist. Therefore, crossing over to adjacent universes should be possible for light. Let's look at the diagram that was presented before.

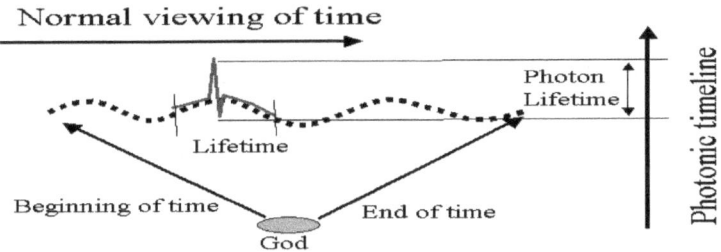

In the diagram I showed that light really had no normal time component but would look like a spire out from the normal timeline, in fact, if we were to recognize the lifetime of a

93

photon, it more correctly is determined as shown above. Beginning and ending in the blink of an eye, but it could last for the equivalent of centuries in what I'm calling lateral time.

Speed of Light Example

In a lateral time world, what would a rocket look like? The answer would be that rocket would look as long as the entire trip the rocket took in time [sort of converted to length]. I brought that up for a reason. What would a rocket going the speed of light look like in "normal" time? You guessed it; it would be as long as the entire trip to a viewer that was stationary.

Vibration

If vibrations are emanations of modification over time, then the rocket could be viewed in lateral time as a solid mass. I know that sounds like a black hole and matter must be being sucked into oblivion, but it's not. Remember, we are looking at a different dimension. To the viewer of lateral Time, all the vibrational motion would now be produced simultaneously. As a particle moved over a distance, it would look like a line drawn across many more lines. Let's explore this just a minute. How do we perceive the vibrational patterns associated with a mass? The answer is that we perceive them as a mass. We cannot see the vibrations and, frankly, we cannot even understand them as vibrations. It is as if our "impressions" of things are associated with viewing matter in lateral time rather than linear time that we experience.

Photonic light would be seen as solid beams like wire and here is another thing to think about. How long would the light beam be visible? ---Da-^{dah}-dah-$_{da}$-da-^{dah}-dah [I'm

94

I mean the image would be there forever in lateral time.

Lateral Time is the Speed of Light

I know we use this "speed of light" thing every day, but now we can define it differently and possibly more accurately.

The speed of light is when vibrations become solid. It is lateral time.

I know you were expecting that absolute zero or minus 473 degrees would make all vibrations stop, but we actually had to go in the other direction.

The speed of light also is the transition from linear time to lateral time.

Once we cross that boundary, time travel, by this description should be possible in forward and reverse by simply injecting yourself into the lateral picture presented in lateral time.

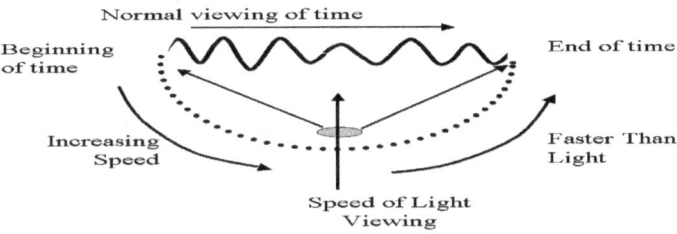

The pervious drawing, hopefully, will sort of explain what I am saying here. The curved arrow on the left shows what happens as you move closer and closer to the speed of light. What you would see is that everything would speed up for you. As you approached the speed of light things would be skittering around faster and faster. Soon there would be a

95

blur and everything would be impossible to understand because you would be seeing all time in an instant.

Beyond the Speed of Light

The blur begins to slow down slowly as you go faster than the speed of light as described by the curved arrow to the right of the drawing. What you notice is that things are going backwards now. Soon you can get to a level that is similar to where you started with respect to the motions of everyone, except that everything is backwards. While in this state, one could gain knowledge of the future that would now be represented as the past. Once the information is retrieved, slowing down to a stop brings you back again to "normal Time" and you have the secret of the future with you. If you noticed, as soon as you go faster than what we call the speed of light, you don't go faster. Instead, you go backwards and that brings us to an important description of time which is certainly needed for any discussion of light and life.

Backward Time

Let me tell you something else about this backward in time vision. You would be seeing an adjacent universe rather than our own. I'm not even sure you would recognize any of the elements of our universe very easily at all. For us, we only need to understand that there is something we can call a Constance of Time.

Constance of Time

Just like all things, light, matter, and life, time CANNOT simply go into infinity and be lost.

If time was only one way, soon we would have no time. Therefore there is an underlying negative to time. We can perceive this negative time as an adjacent universe [some call heaven] where time is completely backwards. If we view this NEW world, their in-waves are our out-waves, their electro-magnetics is our matter, etc. There is a strange linkage between our universes and both are needed for either to survive.

What Do We Learn From Lateral Time?

As we search for a definition of Photons and Light what have we found?

- *Light going backwards and forwards in time can be reconciled in Lateral Time.*

- *The beginning and end of time being together allows us to witness light as a straight function of time.*

- *Light appears to be Matter viewed by lateral time.*

- *The speed of light is when vibrations become solid. It is lateral time.*

- *The speed of light also is the transition from linear time to lateral time.*

Time-Space Dynamo

Like the other dimensional blocks or dynamos, there are three dimensions and each is mutually perpendicular. As shown below left. I know that sounds weird in that I described one of the dimensions as "Backwards" time. To describe this possibility, I turned the diagram slightly to show that forward and backward time are not "Parallel times" The lateral time is shown in all. This represents the very unusual circumstance at the speed of light.

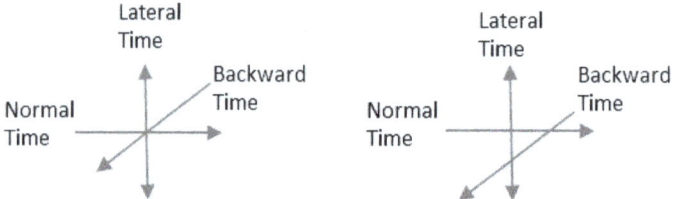

Let me shift the diagram back again and move the adjacent universe timing away and bring up one more thing. The time in each universe is not locked together allowing for what physicists like to call Multiverse. This can be done with 2 universes with times allowed to change independently.

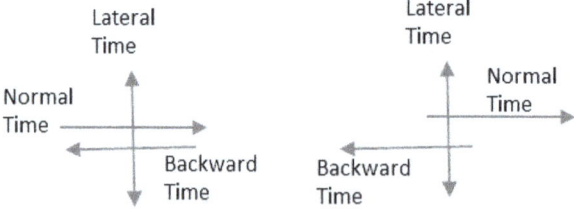

Because the universes are not hard linked to time, one can easily slip to different times during excursions beyond the speed of light. Let me give you an example.

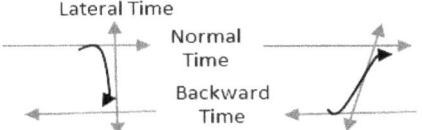

Lateral Time

Normal Time

Backward Time

Notice that the Soul dimension is a window our adjoining universe that can be "jumped" through the time defined as transverse time. As each universe time is not "hard linked" they universe motion through time are different, but the Life dimension holds the return allowing for intrusion and exit to the same entity---even if Apparent time variations occur. I know this is weird especially if you are not comfortable with vibrational matter and radial characterizations of the components of our universe, but hopefully the example helps a little. This excursion was only possible when the soul momentarily left the region we loosely call "ourselves". While no "body" can survive the trip beyond the speed of light, what about almost that speed?

Almost the Speed of Light

Let's slow down to just under the speed of light and look at what Einstein had to say again. Einstein told us and we have later confirmed that as we go close to the speed of light, our mass reduces and time almost stops for us. Let's say two people are together and the same age. One goes traveling at close to the speed of light for 20 years. Neither shaves until the traveler returns. When he returns, the fast guy has not aged a day and the other guy is 20 years older and has a 20-year beard, as shown below. It didn't matter where the traveler went and how he came back so long as he did it close to the speed of light. He could even stop every once in a while to see stuff. For the guy going fast, time almost stops, just like the suspended animation "sleep" method, but this time we simply used what Einstein referred to as "relativity" which forces the speed of light to be constant to all observers. If you are going the speed of light, you must experience light "generated by you" going the speed of light, so you must not experience the time during your travels AT ALL. You are in a different "reality".

Of course they have proved this anomaly with a number of experiments, but again, this only allows the forward dilation of time. If you went on your journey for a thousand years, everyone you knew would have gone and you could never get back to them. That would be a bummer. I know this is a way too brief description of this weird phenomenon, but the fact is that this one way relativistic time travel is not what this book is about either.

Notice I showed that the guy going close to the speed of light does not have to travel in a straight line. Let's see where that takes us.

Instead of going somewhere, the fast guy simply turns in a circle [After all that is the way the Aether travels the speed of light in matter]. It doesn't matter, the same thing happens, in the blink of the fast guy's eye, the slower guy turns old and feeble.

Let's look farther. Let's say the spinning guy turns on a flashlight aiming outward. What would the slow guy see? Each time the flashlight came around in his direction he would be blasted in the face with light. Because of the spinner's speed, that would be all the time.

To the slow man, he would, sort of, see his friend turn into light in this reality.

Now assume that the spinning man was not just moving around in a circle but was spinning in all directions and you see that he becomes a speck of light. The next time you see light just think of it as a spinning man and time travel may become a little closer. Once he got to the speed of light, he would be going backward in time and his out-waves would now be considered in-waves. Let's examine what happened. We know his mass got more and denser as his body became made of different larger and larger atoms as far as the stationary reality was concerned. He got smaller and smaller as space-time changed so that neither made sense to what the stationary person viewed. The high speed guy had no understanding of our reality. To him it was an instant as his partner's beard just appeared. We have to have dimensional descriptions to cover these things that are no longer anomalies.

Tiny

I forgot one more thing in the example that sounds problematic. As the fast man spins faster he gets smaller and heavier. Close to the speed of light, the spinning man is very small and very heavy. Mass tends towards infinity as the object approaches the speed of light. While the spinning man has completely vanished except for the light shining in all directions as he steps past the speed of light. On a linked universe, he is still matter, but with so much gravitational pull he is considered a black hole or portal to the reverse time universe. In the old days we just talked about turning lead into gold----child's play.

Vibrate

Do one more thing for me. Think about what would happen if the man didn't spin around in a circle, but instead, he stayed still and his body vibrated close to the speed of light. He would again stop aging and time would almost stop for him. He would decrease in size and increase in mass---all while he was standing in front of the other man. The vibrator had actually traveled into the adjacent universe, was viewing backwards time as some type of black hole thing.

This has given us more information to define our universe, but we still need more. Let me give you another example I think will help clear up something. In this case we will have an object move away from you and we will look at something called the RED SHIFT that violates the old set of dimensions establishing reality.

Violation of Reality

While I'm mostly talking about light, everything must be considered in this example as it affects our reality. That is because vibration can only be defined in a set time. While length, height, and width defined in a rectangular coordinated world could just sit there vibrating mass has a serious problem. Today we know that things change as vibrational frequencies change. Let's look at some at the stars.

Red Shift and Gold-Dr. Hubble discovered that if objects moved away from us very fast, the colors that emanate from them shift towards a more reddish color and these "RED SHIFTS" are packetized as if blocks of things are all going the same speed from us. To the object moving away the colors didn't change, but when we watch them the colors are all different. The light changed into something else because the number of vibrating cycles had to stay the same as shown in the following graphic had to get slower for us to see it. I think the next example will show you what Dr. Hubble really found.

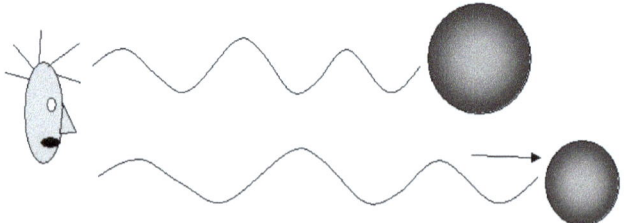

Same number of vibration cycles are created but must be stretched as the object moves away for us to see them.

While this is an accepted characteristic, what it is showing is the universe associated with the object moving away from us and our entire universe stays in sync with 'OUR REALITY" by changing vibrational patterns and vibrational patterns determine matter.

Let me say this again. Time stays in sync by changing matter. The red shift is actually matter being converted to smaller particles in our reality.

What this red-shift means to 'Vibrating" matter is that if we could see the things on the object moving away from us they would **not** be the same things as they would be if there was no relative motion. Let me explain how bizarre this WEL KNOWN phenomenon is.

Lead objects might be gold objects-- or we could "see" radio waves as light-- or any number of things that would make us go crazy. It's best for us not to look.

If the moving away guys could see us, gold objects would be lead and our bodies might be some type of bright blue material or our clear atmosphere may be completely solid. Here is the most important part of this. If a person is moving away from us at near the speed of light, he would gain mass according to Einstein, but he would reduce his mass to lower vibrational elements according to Dr. Hubble. Soon the person you were viewing would be invisible as his vibrations stretched enough to turn him into mostly oxygen rather than mostly water.

Given all that we could say if you go the speed of light, you would turn into Aether in the reality of the viewer.

Let's look at the opposite view. If you are going fast and looking at everything around you and you see a tree. If you are going fast enough, the tree has no width at all in the direction of travel. There is no way the "line" that you see is made of wood and plant cells in your high speed world.

Increasing Velocity

Einstein Confirmed It

If you ever wondered what Einstein's famous formula was all about, let me tell you.

$E=MC^2$ is saying $M= E/C^2$ or <u>Energy and mass are EXACTLY the same thing</u> when there is a differential vibration or speed associated with the speed of light. You cannot have what we call matter at the speed of light so you would be Aether.

By the way! I am not writing Aether like "Aether bunny". I don't write with a lisp. I want to make that clear.

If we assume mass really does exist, we can characterize it in the same way as light by determining its resonant frequency or wavelength. <u>Some believe that mass and light are actually the same thing</u>, but I am not addressing that in this book except in a cursory way for completeness, Right now I am just showing similarity between Photonic wavelengths and matter wavelengths that change depending on something called Time vectoring.

Time Vectoring

I know that some of this is getting hard to understand so I'll try to draw pictures. The matter and electro-magnetic force components of the universe are completely inertial based in that all emanations can be characterized spherically. Unfortunately or fortunately, time seems to act adjacent to the other key elements of matter and force. While forces all must be identified with time and waves must be characterized with time, we now know that this time component is not constant in an absolute. Time is relative. While one could say that means that the structure of the universe would expand or contract depending on the time constant used for the model. It is this dependency on reality that makes time so hard to introduce. Placement of all of these wave nodes starts shifting and confusion ensues. Time, however, seems to react to velocity. While that seems like a self-fulfilling element as velocity only has meaning with time, the thing to recognize in Einstein's model is that time is "Vectorized" as shown next in a picture I drew.

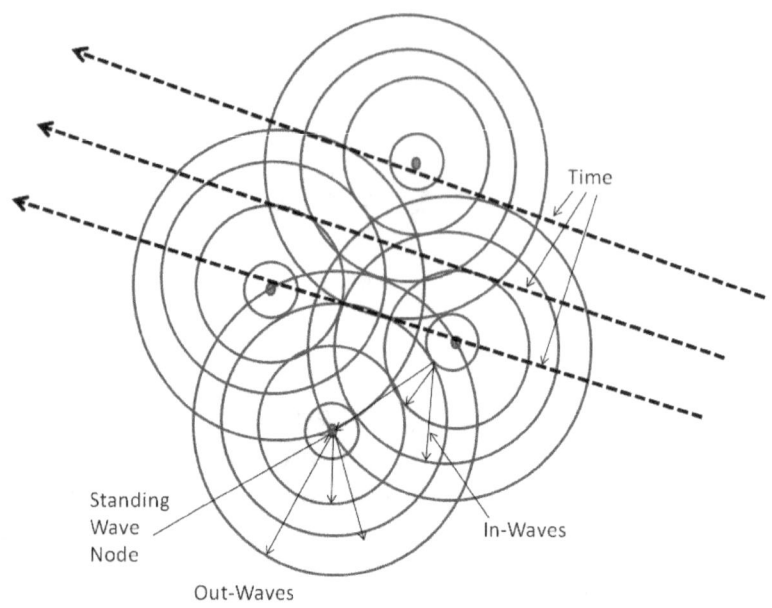

Time

Standing
Wave
Node

In-Waves

Out-Waves

Time stays constant to an observer. Wherever he goes and however fast he goes, **"Time" stays with him**. That is he experiences everything exactly the same if he goes close to the speed of light or if he could possibly not move through space at all. Experimentally, we show that along the axis of a velocity, mass increases. If we could see a table, for instance, the table would get longer in the direction of its velocity vector. NO! NO! NO! In the relative world the table stays the same spatially and the forces [all of the in-waves] react to cause the same forces to the traveler and the objects moving with him. In one way, time can be considered as a rectangular element of the polar universe. Straight-line augmentations of the fabric directed to a vibration. *Because none of us are pulling on this fabric the same, the universe must be characterized as multiple universes all being vector driven by time differential associated with time.* I know you are wondering how we can even have a "stationary reality so let me back up just a little.

108

Lateral Time Dimension

Let's imagine the time dimensional dynamo is a starting point and a framework for vibrations associated with matter, forces and life. In my book on vibrational matter, like all the other dimensions, time has a vibrational aspect. What I want to briefly discuss is dilation of time which has no visible change to us. You can only witness it when you are outside the confines of it. One hard part of understanding this dimensional block is that the time dimensional string is closed just like all the rest. What that means is that is repeats itself. It may do this repeating thing thousands of times and we would not normally know it. While time can be considered constant, we actually base our perception of time on the cyclic nature of our brain function so each person has his own base time that is unique to him. A one second thing, to me, is completely different to a one second thing to a bird, for instance. We can say time is dilated for us when compared to perceptions of the bird.

Closed Strings And Time Dilation

If the components that make up the frequency and distance don't match the closed environment, <u>the distance environment must be changed to insure that the vibrations DO match.</u> We can call this a time dilation effect defined by vibration. If we could see the time dilation, we would find that everything gets larger. The graphic below illustrates the expansion associated with time dilation.

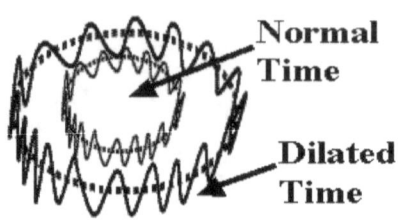

One way to recognize time dilation is by using the dimension I called "Lateral Time". From this "vantage point" one could see all time being stretched but those locked in a recognized reality, those changes cannot be registered.

Seeing the Future

Like it or not, it seems that we can experience several predecessor and future lives that are unveiled in hypnosis therapy. This is well documented. To, possibly, explain this, let me revert back to an old philosopher Immanuel Kant and several other similar teachers. They surmised that people could be represented as hanging entities in an empty environment. The people might be hanging from a floor or ceiling as shown below left.

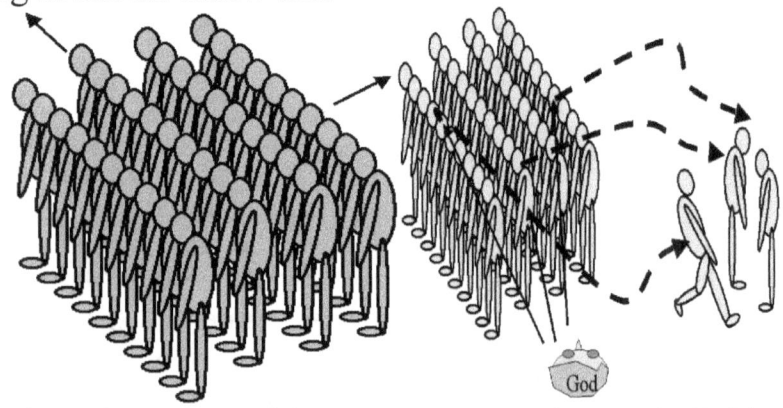

He believed God would represent a reality to individuals such that they would live in a world outlined by the details put into their consciousness. This is represented in the diagram above right. Generally speaking, 2 or more people

110

would have similar inputs and they would react to the "introduced" environment. Now that I have gotten you confused again, let's say "God" introduced several environments into a person simultaneously. One might be in 10BC; one might be in the 15th century; and one might even be in the future all at the same time to the viewer of lateral time. The consciousness of the person might be able to remember all the events because he was a part of all the events. He could "remember" the future. The problem is that in this philosophical world, nothing is actually real. Like the previous lateral views, the one shown describes many "lifetimes" remembered and experienced.

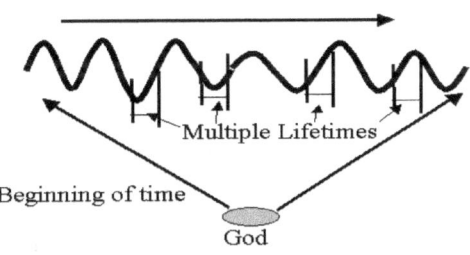

Let's Make a Reality

While this lateral viewing thing seems to sort of work with God seeing the beginning of time and all, what if we could view the same lateral timeline and what if each one of the different lifetimes that are viewed are different people or the consciousness of the same person being "recycled" over time? What I mean here is what we typically call reincarnation. If one could interpret lateral timing, one would be able to see the future and the past as if it were all accomplished instantly. Just think about it!

Speed of Light Example

That was the easy description. In a lateral time world, what would a rocket look like? The answer would be that rocket would look as long as the entire trip the rocket took in time [sort of converted to length]. I brought that up for a reason. What would a rocket going the speed of light look like in "normal" time? You guessed it; it would be as long as the entire trip to a viewer that was stationary.

Getting Older Example

Let's look deeper. If you could see someone who was viewing you in lateral time, how old would they get as you aged? Of course they would not age a day because they could see your entire life as an instant. If you go the speed of light, how old do you get with respect to those not going with you. The answer is that you would not age.

Vibration

Now for the hard one, if everything is really made up of vibrations, as I have presented in the book on vibrational matter and this one, what would everything look like when viewed in lateral time? Hurry up; the clock is ticking. ----- Come on! If vibrations are emanations of modification over time, the answer would have to be a solid mass? I know that sounds like a black hole and matter must be being sucked into oblivion, but it's not. Remember we are looking at a different dimension. To the viewer of lateral Time, all the vibrational motion would now be produced simultaneously. As a particle moved over a distance, it would look like a line drawn across many more lines. Let's explore this just a minute. How do we perceive the vibrational patterns associated with a mass. The answer is that we perceive them as a mass. We cannot see the vibrations and, frankly, we cannot even understand them as vibrations. It is as if our "impressions" of things are associated with viewing matter in lateral time rather than linear time that we experience.

Photonic light would be seen as solid beams like wire and here is another thing to think about. How long would the light beam be visible? Da-^{dah}-dah-$_{da}$-da-^{dah}-dah [I'm humming the Jeopardy tune.]-----"What is forever?" is the correct question.-----I mean the image would be there forever in lateral time.

Backward Time Dimension

Our "connected universe" converts out-waves from our universe to in-waves. The major difference in the 2 is that one is going backwards in time. A silly way to show this is that in-waves are leaving our universe just like out-waves, but they are going backwards in time. As our out-waves leave, they instantly become in-waves that produce "forces" needed to run that universe while their out-waves produce forces to hold our universe together. Without the 2, there would be no universe. Unfortunately a universe would still be nothing without life which also appears just as waves of nothingness.

No matter how much of the vibrating nothingness you have, there still is nothing.

Life Dynamo

The Self

To define life we have no real scientific group trying to define the life energy, but let me give it a try, knowing the similarities of the other 2 dimensional dynamos. Hopefully from the previous chart of equations you can saw the similarity in structure to Electricity and Aether as both act as a potential or as a capacity. This established the Energy of what we can call the SELF. It resides in what we call reality as soon as it begins to vibrate. Some call this the essence of cognizant being. Some call this the Consciousness or the conscious mind. It is the dimension that experiences reality as we normally view it.

The Subconscious

The second part of life and perhaps the most important part is the kinetic part we can call the subconscious or the Soul. Like electro-magnetic and mass-gravity waves, the soul is an existing characteristic of life, but it is so much more than that. Great groups of "soul-particles" resonate in a similar way and establish <u>consistency of life</u>. We call this reality. Here is the neat thing to change reality, all you have to do is to increase your vibration frequency and go BEYOND it. All that "Think and Grow Rich", the "Power of Positive Thinking", and "enhancing one's Faith" all are actually turning out to be truth. To be "self-Actualized" for instance, allows you to help others because you can help them increase their "vibration frequency. I know it sounds bizarre, but we see it all the time. We just don't recognize it. If you ever wondered how someone traveling at close to the speed of light would have no sense of time in this reality and would not age in this reality, it is simple he is not in this reality as his soul is vibrating too quickly.

Spirit

That brings us to the last "life" dimension we can call spirit. Just like pure magnetism and pure gravity, this dimension actually allows for transfer between universes. In this case souls can be transferred or they can re-enter this universe. While our "self" would be destroyed going the speed of light needed to cross universes. Things like time travel start making sense as a SOUL could experience a different time in a "backward time universe" a soul could even enter this universe in a different time. All sort s of things begin to make more sense. Have you ever wondered why the Bible indicated the creator was actually 3 entities and that we were made in his IMAGE as 3 entities?--- never mind the analog, we are now finding out that some of this religion stuff is real.

By similarity you can see that I have interpreted the Photonic, Nuclear and Life transfer energies as being similar and <u>tied to the speed of light</u>. We can test the nuclear action and can generally test the photonic energy levels, but to my knowledge, no one has the insight to finalize energy transfers that would be required for a trans-universe conversion. Therefore, these equations must stay notional until one can attempt a transfer.

It might be recognized that mutually perpendicular systems are stable and complete stability may be the mark of a universe.

The universe doesn't exist without life, but let's expand on this LIFE thing.

Let me say that life is NOT DNA. Putting amino acids in a double helix crystal will not make life no matter how much lightning you hit it with.

Adenine, thiamine, cytosine, and guanine somehow make your cells reproduce and build stomachs and things, for sure, but life is different than a stomach. Life is a true dimension. Like all dimensions, this one is vibrational and lives are connected to this universe in some way we cannot easily understand. The life/Self dimension might be thought of as a "Life Cloud" containing the living. We move around, hike, learn, love, and kill, but all the while there is some "life" that is a connection to this universe. The actions of life are similar to the other dimensions. Each individual's life is vibrationally quantized to this life dimensional string just like the other vibrational dimensions.

Just like the Aetheric dimension that builds individual chemicals, this life thing builds individual- individuals.

I think I need to stop explaining and just get to examples and ancient textual verification on most of the elements of this last dynamo. All I can tell you for sure is that the universe doesn't exist without life and the more aware a being is of its life, the higher would be its vibrational frequency. For instance, bacteria would be a low frequency while humans would be represented by a much higher frequency and life-energy. To try to understand life [the dimension of life] let's look at some of the ancient texts and see if we can glean information that will be useful to us. I have selected a small sampling from the ancient Gnostics, and Essene to help out here. Both of these groups tried to understand the extremely

116

ancient texts of their day. The Essene tried their best to right down the exact meanings of the ancient words, while the Gnostics tried to put the details into a story so that the words could be understood better. Our first stop is to the "Book of Giants".

"Book of Giants"

As you might guess, this book is about ancient giants that ruled over mankind many thousands of years ago. From the Book of Giants [part of the 'Essene's" Dead Sea Scrolls] we read about angels who came from another universe [heaven] to help mankind. Unfortunately, the same old story we have heard a hundred times happened during this ancient time. The angels sort of began to think of themselves as Gods and let's see what happened.

*For they [**200 inhabitants of our linked universe who came to teach mankind.**] knew the secrets of heaven and sin was great in the Earth because of their experiments. They made mistakes and they killed many animals and people.* [**What we see is that these "intruders" tried to make or modify life itself. This ended up killing a bunch of people and animals. I know we are doing the same thing today, but let's continue.**]

They were with women and they begat giants. [**If you can't change something one way there is always sex.**]

They selected two hundred donkeys, two hundred asses, two hundred rams of the flock, two hundred goats, two hundred other beasts of the field. From every animal, and from every type of human was taken its seed for mixed sex. After a time they defiled the animals and people and begot giants, monsters, and dragons. God saw all that they begot, and, behold, all the Earth was corrupted with their blood and by the hand of man. [**I think you can see that these giants**

weren't creating ANY life, they were simply mixing DNA to bastardize God's creativity in making LIFE.

They were brought food which did not suffice for them and they turned on mankind. They began to hunger and they were seeking to devour many animals and people. The people ran to a safe place but the monsters and dragons attacked it. Man's flesh was eaten by all the giants, monsters, and dragons. The monsters thought that they would be saved and they would arise after death, but it was not so because they were lacking in true knowledge of heaven and because they were of the Earth. They grew corrupt and did not worship the almighty God. They were considering separation of Giants from the angels upon the Earth, but to no avail. In the end they perished and died because they caused great corruption in the Earth and because they tormented the Earth. Suffice to say they were tormented after death.

What we find is really smart humans living during very ancient times. While they were smart, they didn't know what life was. These people were creating all types of animals at this ancient time. The Bible called the animals "Unclean or abominations" and we are just now relearning how to do similar types of animal creations. There is no doubt that the ancient geneticists were more knowledgeable about life, but they still had NO IDEA what it was. Let's read further. This time we will read again from one of the Dead Sea Scrolls.

"Book of Secrets"

To give you a feeling of how different life is than DNA we have to bring up the "Book of Secrets". The "Book of Secrets" found with the Dead Sea Scrolls provides a strong warning about the use of "secrets of God". The book simply says that if we try to misuse the secrets of life, the same thing will happen to mankind that had happened before. The earth would be destroyed again. This destruction would not be by direct intervention of God, but because we, as humans, don't understand what we are doing as we manipulate "Nature". Of the secret elements indicated in the text, it seems that the "manipulation of creation" or trying to create life is the worst. This probably references the genetic manipulation and transmutation or one material into another [Alchemy]. By all accounts, the Ancient people living during the time of Adam employed both of these things before and immediately after the worldwide flood at the end of the Pleistocene Age. Here are the major elements of what has been pieced together of the "Book of Secrets"; judge for yourself.

If it makes you fearful, you read it correctly.

Those who would penetrate the origins of knowledge, along with those who hold fast to the wonderful mysteries of magic; --Those who walk in simplicity as well as those who are devious in every activity of the deeds of humanity; those with a stiff neck, and all the mass of the Gentiles, [Gentiles were all people who were not considered "Chosen Ones" like the Jews and other pure descendants from Noah.]

--With this I beseech your attention. All of the secrets of sin and magic were known but they [The preflood humans, including the Nephilim] did not know the secret of the way things are nor did they understand the things of old. [They didn't understand what LIFE really was.]

They did not know what would come upon them, so they did not rescue themselves without the secret of the way things are. It is true that all the peoples reject evil, yet it advances in all of them. Who wants his money to be stolen by a wicked man, but where are the people that have not robbed the wealth of others? What shall we call man who will call no one on earth wise or righteous? It is not a human possession to act on wisdom. It is not possible because wisdom is hidden except for the wisdom of cunning evil, and the schemes of Belial who modified creation, a thing that ought never to be done again, except by the command of his Maker. [This is the important part coming up]. Only God has the wisdom to modify creation. Belial [one of those from our linked universe] modified creation and it should NEVER be done again. [This seems to include genetic manipulation, alchemy, and all the other magic areas.]

Consider the soothsayers, those teachers of sin and magic. Do not regard your foolishness, for the vision is sealed up from you, and you have not properly understood the eternal mysteries. [Just in case you didn't get it the first time, it is stated again. Manipulation of nature and Magic cannot be understood by man. It is foolish to try to use these things for good. I know we are using genetic engineering today to help many things, but this is saying the apparent help will not last.]

121

You have not become wise in understanding my secrets; for _you have not properly understood the origin of Wisdom._
[Here is all one has to do to be able to correctly use "magic". In order to understand how to manipulate nature in a good way, you must understand how it came into being. Unfortunately you cannot because you didn't exist when this Wisdom" was established.]

So if Life is one of our dimensions and we don't understand what it is, what are we to do? Well, it looks like we are getting much closer to understanding this elusive dimensional quality of our universe.

Self Dimension

No one completely knows the meaning of life no matter how many people say that they have found it. The examples above were talking about ancient people who lived longer and had more historical understanding of how animals work, but EVEN they did not know what Life was. While the exact details of what life is, we can be pretty sure that it is CONNECTED to the universe. If matter cannot be created or destroyed, we can assume that LIFE has the same characteristic. I can tell that I losing you. You're thinking that when someone dies, life must be lost, but life may be reborn somewhere else. You are thinking that the number of people continuously increases over time, but in fact, if we look at the people living on the earth over the 40 thousand years from Adam on, we would find that there have been many peaks over the ages with greater numbers of people than are in this world today.

Killed off by war, famine, disease, and age, human Life may even be recycled so to speak. That does not mean that a person becomes a bug in our next life, so don't worry that you stepped on grandma.

What it means is that there is a LIFE energy that is REQUIRED to sustain the universe and it is all around us either as living entities or similar things that God created before ANY life-form was able to LIVE. Life is the first dimension of this last, but most important dimensional dynamo that is required to be here so that our universe will exist as a reality. Like the other previously described dimensions, Life can be considered a vibration.

By now you should enjoy the knowledge that the vibration I have been talking about is not like sound waves or light beams vibrating to bring information to our brain. It is the one-dimensional essence that determines and establishes our universe and our very being.

Modifications of this vibration change the characteristics of life itself. It can become more dynamic or less expansive. Higher frequency vibration components may live longer or the life function may be enhanced. This is not consciousness here. Consciousness is something entirely different than life.

Self

The life dimension is generally characterized as the "self". To describe this dimension of life, let us call it the potential for life, similar to the electrical potential and the Fermionic potential we studied previously. An example of this dimension can be easily rendered. Let's say a woman with ample bosoms walks by. While you want to think about something else, your "Self" is the part that is drawn in by the objects of stored fat on a woman. The desire and mood make absolutely no sense, but self can be characterized as internalization of reality and recognized as sensing self, survival, and sex. The book of "Proverbs" has something called the 7 carnal sins and the book of "Galatians" has something called the sins of the flesh, and later these were compiled as the 7 deadly sins. These really describe the self. While they don't seem bad, what these things do is weaken control of the most important part of life which we call the soul. Here is the description of self.

124

Proverbs 6:16-19- *These six things doth the LORD hate: yea, seven are an abomination unto him:*

- ***Pride*** *[in yourself]*
- *Lying [to protect self]*
- *Killing[to protect self]*
- *Wicked imagining*
- *Making mischief*
- *False witness*

Many of these don't seem like horrible things, but they drive us into having "self" control" us rather than the soul.

Galatians 5:19-21-Now the works of the flesh are evident, which are:

- *Fornication/sex,*
- *lewdness,*
- *Idolatry/sorcery*
- *Hatred*
- *Envy/ jealousies,*
- *selfish ambition,*
- *murder,*
- *drunkenness*

Those who practice them will not inherit the kingdom of God.

Like the first set, indicated, we control our universe with our soul, <u>we satisfy our self by ignoring the soul.</u>

Later Compilation

These were all, sort of compiled into a single theme as indicated below:

- gluttony
- lust
- greed
- pride
- Sorrow
- hatred
- vanity
- laziness

I know you would not think that feeling sorry for yourself would be a horrible thing, but what was being said has relevance to us today and it is truly horrible. I think the Taoists said it best.

They indicated that to live a life that is meaningful one must think of himself as an empty vessel. [Forget about self to experience and gain TRUE life].

Later this was called self-Actualization. The Bible had a different way of expressing it, that I described earlier.

Mark 8:31-9:6- For <u>whoever wants to save his life will lose it</u>, but whoever loses his life for me and for the gospel will save it. <u>What good is it for a man to gain the whole world, yet forfeit his soul?</u> **[Forget about self to experience and gain TRUE life]**

With that let's look at the soul or the dimension I call "consciousness". It may be better defined as consciousness of the "unity of the souls controlling our universe" over consciousness of "self". With the new Anthropic view of our universe, knowing how critical the dimensions of life are to reality, we must reassert the foundations concerning what this universe is actually defined by.

- *There is a reason a small woman can lift a car off her husband and her bones don't break.*
- *There is a reason Moses and the Egyptian magicians could turn a stick into a snake.*
- *There is a reason reincarnation is not only possible but a requirement of existence.*
- *There is a reason Elijah, Peter, Elisha, and others could walk on top of water without falling in.*
- *There is a reason Jesus told his followers "With the faith of a grain of mustard seed one can change the place where a mountain stands." [He was not saying mustard-seeds had all this power, he was saying if we let ourselves allow our souls to take control, we can EVEN change what we typically consider reality.]*
- *There is a reason why an idiot savant can play a piano without lessons.*

Let's look at this very critical dimension.

Soul Dimension

This is a little confusing as the Life dimension contains all the self-preservation, sexual urges, procreation desire, survival, and gratification which is somewhere between the subconscious and the debase conscious thoughts to gain comfort, greed, selfishness, and all the other Carnal desires. The functions are machine-like because they are simply the person or mind's reaction to stimuli totally set in this reality. The soul portion of life is freedom FROM our carnal nature. The more powerful this portion of life is the more interaction one can have in controlling his destiny and the destiny of those around him.

The notional energy equation for the Soul dimension is given again below. As you recall, there is a certain consciousness factor that is directly proportional to the energy of the system and as the vibrational frequency increases, the soul energy increases and the "self" decreases. What we will find is that modification of the vibration of this dimensional component can be done automatically, inadvertently, dynamically, and even haphazardly.

Like the other dimensions if an entity is totally soul [like total magnetism and total gravity] there is a window into the spiritual world of our adjacent universe. As self and soul are always battling for dominance, the self usually wins, but since the Athropic scientists have begun opening our eyes we are now asking the following:

Does our "soul" add to or take away from our universe? If we did not think about something would it disappear?

This type of question has haunted philosophers for ages along with the following even more curious question.

When you die does your subconscious die?

If you die, your subconscious energy cannot halt. The items around us allow our souls to synchronize with the universe so to speak. When our body dies, our soul has more "freedom".

We become part of the universe and add to it with a vibrational union.

To find out a portion of what this important dimension is, we must first separate the ID and EGO. Life does not necessarily mean consciously aware life. It is only life. Grass, for instance, is alive, but, most likely not conscious, an amoeba lives, but acts mechanically and it seems to have no consciousness. On and on we could go separating living and consciousness. The reason we can separate them is that they are very different. Like the relationship of electricity and magnetism, self and soul are very much related. Let me give you an example that everyone can relate to. The great researcher, Abraham Maslow, discussed the massive changes that take place in a person who becomes self-actualized [having the soul portion take more control]. The more a person gets in-tune with his own consciousness, the more he can and must help others. By the time he is totally aware or self-actualized, he can be the most help to other LIVES by totally disregarding self and focusing on what can be done. This increase in consciousness and its closer involvement with life shows the close union of these two dimensional strings.

128

Vibration And The Brain

I know you are struggling with the concept that if you become less conscious, the universe is somehow affected, but the idea that our consciousness is interconnected with the universe is not a new one. Even the concept that vibration controls the level of consciousness is not new. While the concept isn't new, new studies in vibrations and the affects to or by the brain are popping up just about every day. Let's look at the list of changes in the brain caused or described vibrationally. I'm talking about the alpha, beta, delta, theta and gamma brainwaves and how they affect us. To test and determine what vibrations affected the brain in what ways, there have been 2 major methods deployed currently. Both of these are audio entrapment methods. That is they allow the brain to interpret much lower frequencies than those actually transmitted.

Two Testing Methods

The first method binaural interpretation is accomplished by sending slightly different frequencies in each ear. The brain picks up on this difference and uses the difference in its normal excitation.

The second method is called modulation. Similar to the other, this method simply modulates any music or tone very slightly. The brain senses the modulation and interprets that as its control function.

Certainly there have been blinking lights, mechanical vibration, and magnetic modulation methods, but the audio methods are extremely easy to accomplish. The Infratonic Qui Gong Machine, for instance, was developed out of scientific research in Beijing China which studied natural healers and found that most powerful healers were able to emit a strong infrasonic (low frequency sound) signal from their hands. The sound emitted from average individuals was only a hundredth as strong. The "Infratonic", is now used by 1% of all doctors in the United States and it is believed that it also is an audio modulating device. A Dr. Keely designed something called the Krell Helmet that relied on electromagnetic fields generated in the helmet. I don't know how successful this machine was, but it illustrates the point that everyone is trying to artificially excite different levels of consciousness and some are beginning to get success.

Not everyone has gained success as can be illustrated with something called Sphincter Resonance. In the 1960s, somebody discovered the resonating frequency of the sphincter. Presumably, this team created a device later called an "Anal Sphincter Resonator". It was, supposedly, kind of like a musical organ. The idea was to intensify the "suspense" in movies whenever "Danger" was about to be portrayed. BACKFIRE and more BACKFIRE. Apparently it caused the entire audience to soil themselves. The specific group of tones generated by this contraption has been referred to as a 'Brown Note' for some reason that I am not going into at this time. The specific notes have been lost over time, so I'm sure one of these mishaps will occur again in the future, which brings us to this table.

Type	Freq. (Hz)	Normal Reactions
Epsilon	<0.5	Extraordinary states of consciousness, High states of meditation, Ecstatic states of consciousness, High-level inspiration states, Spiritual insight, **Out-of-body experiences, Suspended animation.**
Delta	0.2 to 4 Hz	Confusion, boosting intuition, Deep sleep, Lucid dreaming, Increased immune functions, Hypnosis, Anti-aging, Increased intuition, Inner being & personal growth, Trauma recovery, Near death experience, Blissful "being" state
Theta	4 – 7 Hz	Arousal, Deep relaxation, Increased memory, Creativity, Hypnagogic state, Access to subconscious images, Reduced blood pressure, Profound inner peace, emotional healing, Inner wisdom, Faith, psychic abilities, Twilight sleep learning, Vivid mental imagery, Military remote viewing
Alpha	8 – 12 Hz	Relaxation, Meditation, Light relaxation, Positive thinking, Creative problem solving, Mood elevation, Stress reduction, Intuitive insights, Daydreams, Calm, relaxed, Lucid mental states, Tranquility, Detachment
Beta	12 – 30 Hz	Alertness, Anxious thinking, Active concentration Analytical problem solving, Judgment, Decision making, Increased mental ability, Focus, Good for absorbing information passively, Treating Hyperactivity, Sensorimotor Rhythm, Outer awareness, Arousal, Dendrite growth,
Gamma	30 – 100 +	Motor functions heightened, Boosted memory, Enhanced perception of reality, Binding of all senses, Increased compassion, High-level information processing, Natural antidepressant, Positive thoughts, Higher energy levels, Decision making in a fear situation, Muscle tension, Release of growth hormone, muscles, Recovery from injuries, Rejuvenation effects

131

The previous table shows some of the findings of researchers trying to modify the conscious mind externally. While there are specific frequencies that cause each of these effects, I'm not going to go into that detail in this book. My main objective is to show that vibrational fields greatly affect consciousness. Don't worry that these frequencies are so very low in frequency comparted to the super high frequencies needed to convert matter into different substances. We don't know why these low frequencies "activate" various parts of our brain and consciousness. These are simply observations. While the military is experimenting with broadcasting subsonic waves to affect brainwaves and enhance the Delta levels [to confuse and put fear in an enemy], many are now trying to tap into meditative states and learning ability by transmitting the 40 Hertz level. What we are finding is that simple stereo speakers may be the best tool to introduce these GAMMA enhancers. A 200-hertz tone is shot into one ear and a 240-hertz sound is transmitted into the other ear. The brain gets both of these frequencies and tries to mix them together to understand the sound. When the 2 frequencies a beat together, the output becomes the difference or 40-Hertz and the brain begins to learn faster.

Brain Vibrations

The reason I brought up this effect is that it describes how sensitive our consciousness is to vibration. The reason is simple. Consciousness is a vibrational dimension and brainwave studies are not the only way to recognize the vibrational characteristics. A second way is something called "chakra" by Buddhists so let's look at some of these mystical things. According to the believers and the testing skeptics there are at least seven of these chakras or levels of consciousness. As someone increases his chakra, we are told he feels a vibration all around and inside.

They are sort of represented by the diagram below.
- Root, Chakra--- "consciousness of Survival"
- Sacral Chakra --- "Consciousness of Sex"
- Solar Plexus Chakra --- "Consciousness of Self"
- Heart Chakra --- "Consciousness of love"
- Throat, Chakra --- "Consciousness of the truth"
- Third Eye Chakra --"Consciousness of inner being"
- Crown, Chakra ---"Consciousness of the spirit world"

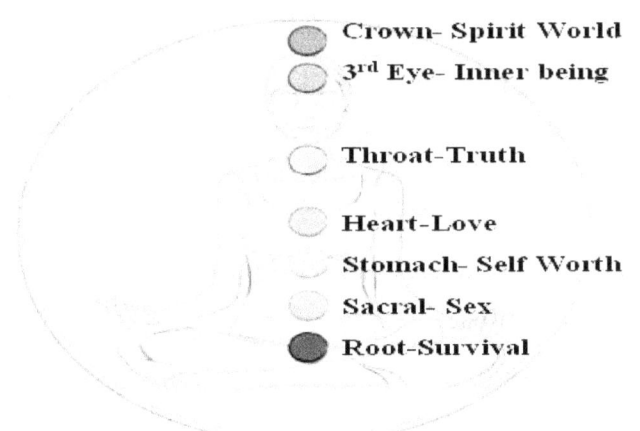

Crown- Spirit World
3rd Eye- Inner being

Throat-Truth

Heart-Love
Stomach- Self Worth
Sacral- Sex
Root-Survival

I know it sounds like I'm some guru from India talking about chakras, but it is a convenient way to discuss this dimensional component so I'm going to continue. I'm not putting on the towel on my head, but I may hum a little as I write this section.

Remember what this is all about is the rejection of the Self, Survival, Sex debased thoughts that restrict knowledge, wisdom, and enhanced control over our environment to essential do what the Bible recommend.

If you want to live you must die [so to speak].

Let's say a girl witnesses her father being pinned under a car. She immediately rejects self, sex, and survival and this allows her to "move mountains"--- or a car in this situation. As soon as she realizes what she did, reality goes back to normal.

An idiot savant who is not as in touch with this reality taps into the "universal knowledge of music. His brain and muscle memory all are modified and he immediately "knows" how to play. Luckily, his rejection of reality continues as his body catches up to the new found knowledge.

Elisha walks on water by doing the same thing as his body no longer is heavy enough to sink.

John Hutchison's experiment making a bowling ball rise off the table is the same, but he simple introduce ultra-high beat frequencies on the scene with his equipment to momentarily change reality in the "focused" area.

Bowling Ball Flinging

While some of the things the Canadian Researcher, John Hutchison, has accomplished in the last few years are remarkable, let me just show you a couple of the frames from one experiment in the presence of a powerful and strange field of electromagnetic waves. First pliers were picked up and yanked out of the view, but then, a bowling ball rises off the table until the ultrahigh frequency vibrations were halted and the bowling ball resumed its normal heaviness in our reality.

I must say that John typically has trouble duplicating his results, but they are still something special. Like another researcher who had similar results, John Keely, they did not understand how very important the vibrational level of the observer was on their results. With that as an introduction, let me get into more about Participatory Anthropics.

Anthropics

This brings us back to the question I indicated before, but maybe this time it won't sound as stupid. If a tree falls in the woods, does it make a sound? The answer according to Einstein and many others is that it not only does not make a sound----- it simply does not exist.

The universe is "brought to <u>life</u>" by the third dynamo of our universe.

Without life, there could be vibrations, but ---so what!!! People, to be more precise—souls, combine together to "create" what we can call the carnal world. This is the world we generate in our subconscious to be enjoyed by our consciousness. The theory of Anthropics allows the combined "resonances of the soul portion of life to makes us understand what we call reality. Before we can understand how vibrations turn into a door, or friction, or photons, we need to understand the last of the three dynamos that make up our universe. I call it the Life Dynamo of dimensions. It is essentially made up exactly like the other 2, but let me sort of tell you the similarities rather than building more equations.

Relativity Helped with Anthropics

This is actually the first discussions of Life dimensions, but it may help a little bit here with time, so let me at least bring it up. We need to add something else to the model or all of the relativity, and Anthropics have issues. Hopefully, my drawing below, sort of, shows how all this fits. While Matter and forces are affected by time, <u>the collective consciousness cannot sense it</u>, but it must exist to address relativity, so it is perpendicular to both Mass and Force and perpendicular to the effects of time. I drew it 2 ways to, hopefully, describe it a little better. Some may call this the soul or soul collective.

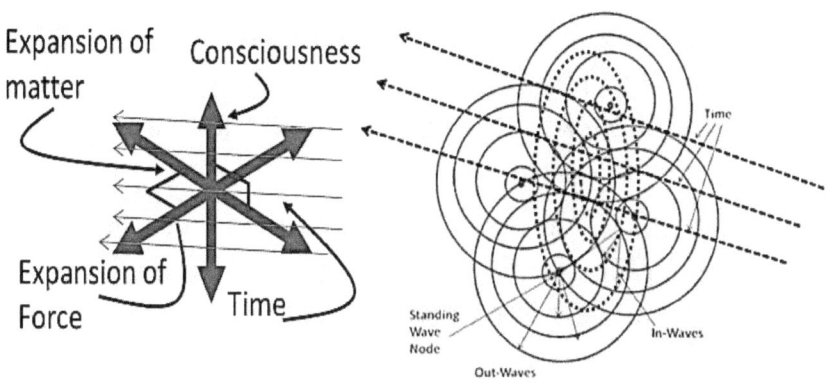

Soul

<u>The meaning of life is not 42</u> as indicated in a popular science fiction film. Instead, it is a complex variant of our universal existence. It is said that if a tree falls in the woods and no one was around, the incidence simply does not

happen. Without a cognizant observe, the universe modifications cannot take place. One way to tackle this subject is connecting life to time. That is, time without life or vice versa means little. More precisely, connecting time and the conscious mind [soul] must be done to characterize the universe that changes by perspective in Einstein's relativity theory. They are both undeniably attached and work as a single unit. That is not to say the soul is time because cognition seems to be "timeless" or accomplished adjacent to time. This is saying that the conscious mind is projected into the universe perpendicular to time so it does not place restrictions to time.

Because the vector is perpendicular to the various velocity vectors and abstract to the in and out waves, this soul dimension is not characterized in the normal sense. You could say it "usually" doesn't exist. I would simply notice that the soul has complete freedom with respect to time. Because the soul has this freedom, people have been able to sense the future and see the past. People have been able to lift automobiles off people who were trapped without crushing their own bones. People have been able to walk on water and turn sticks into snakes and all the rest identified in our most holy historical references.

Now I will ask. If you see 700nm photons and someone else sees 700nm photons do you see the same thing? Common sense would suggest that how the brain processes the various wavelengths of electromagnetic pulses would have to be different, but if consciousnesses are linked, when I see red, others see the same thing. Don't let people come up to you and tell you that you simply define red one way and they define red another so that we can talk about wavelengths in a similar way--------

We are all attuned to this changing reality and we are responsible, in a small way for changing our perception of reality and the color red.

It doesn't matter how fast we go or how strange someone is, Red is not just defined, and it is established by our collective consciousnesses.

Soul Enhancement Levels

Survival and Sex

Whether we admit it or not, every one of us battles these things every day. The most basic root chakra is triggered when you get hungry and the sex one, well; it shows up from time to time. I'm sure you recognize that these components of our life are, pretty much, uncontrollable and they should never be associated with conscious control. In fact, these "feelings" run parallel to conscious thought always ready to break or modify our consciousness. I only brought them up again because some identify them as consciousness levels. A rule of thumb might be if it is associated directly with pleasure, pain, desire, or self-preservation, it is not part of this dimension of your life. Some self-preservation elements such as moving your hand away from a burning flame is identified as autonomous. I would contend that other things are autonomous as well.

Self, Love, and Truth

If you can get past those you start considering self-worth and even love. Most people spend most of the time going back and forth between the lower 2 and the others. It's sort of like a yo-yo or they interact simultaneously. Slow vibrations, higher vibrations, slower vibrations, higher vibrations and still higher vibrations when someone pays me a compliment or a figure out why a light bulb turns on in the refrigerator or I answer one of those "Are You Smarter Than a 5th Grader" questions. Answering one of the "Jeopardy" questions correctly might even get you into the heart chakra.

The heart chakra is a vibrational level associated with "Real" love rather than the sexual one. I believe that an ameba and a tree have no capability of love, but they are still alive. Sometimes this "heart" level happens naturally for a brief time and you can't seem to even think about yourself at all. If you work at it you can get to this level periodically throughout a day and look at people with true look. The Bible called it loving people as you love yourself. Anyway, most just think they get into love and it is more basic. That type of love puts you below the stomach again. Anyway you must conquer love to some level before you can even get to a point that looks for "real truth rather than "Vain truth" that we usually accept or desire. Real truth is a truth that is truth no matter how it affects the event or who thinks it. It is usually not a popular truth or even the one you would hope for. It simply is. While it seems that this would be easy to understand and use. People almost never are tuned to this type of consciousness so they accept what they believe rather than what they should believe. Let's say you get an openness to understand real truth, there are still 2 more levels of consciousness to be considered. The next is called the third eye.

The Third Eye

The third eye is derived from a little gland in the brain called the pineal "pinecone" gland. The pineal has no apparent use, but it is thought to have been used by our brains at one time. After all, the gland didn't just grow there for no reason, so let's travel back to the Tower of Babel.

141

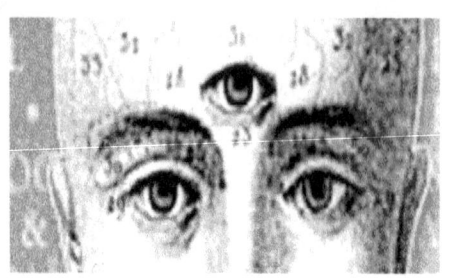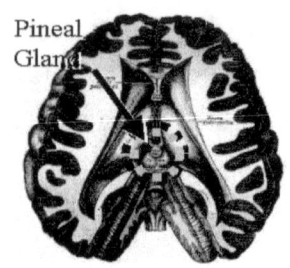

Pineal Gland

Tower of Babel

Most of you know something about this huge Tower that King Nimrod had built about 6 thousand years ago, but many may not know about the huge wars that were written about during this time and how 1/3 of all the people on earth were killed as a result of this war. Called the Bharata War in India with the ending called Zep-Tepi [new beginning] by the Egyptians, we know this was a horrible time on the earth. I could go into what happened when the Tower of Babel was destroyed thousands of years ago here and how our brains lost most of their capability according to many ancient texts and how this brain loss was probably from some DNA mutations. I could also bring out the unusual fact that our current brain size is smaller than our earlier cousins, Neanderthal. While that fact is well known, what is not recognized is that this reduction in brain size shows that our brains began atrophying from disuse about 6 thousand years ago. I could bring out the fact that the entire world was plunged into some type of Stone Age re-insurgence 5 to 6 thousand years ago and people seemed to become dumb as stumps for a while. The Biblical book of Jasher simply tells us that 1/3 of the people died, 1/3 of the people became like apes and 1/3 of the people were dispersed to places around the world because they could only talk to their close relatives. We can imagine that before this brain reducing started, we could do many things with our bigger brain we cannot do today. We can imagine that the pineal gland, prior

142

to whatever happened 6 thousand years ago also was larger and might have been used by our ancestors. I could also bring up many other things that would make you wonder if the pineal gland used to allow us to do many things in the past, but I won't. Instead, let me tell you what this tiny, pea-shaped gland does.

Pineal Glands in many non-mammalian vertebrates have a strong resemblance to the photoreceptor cells of the eye. Some evolutionary biologists believe that the pineal cells share a common were the ancestor to retina cells in the eye.

In some animals exposure to light of this gland can change the animal's biorhythm.

Some early vertebrate fossil skulls have a pineal opening so that it probably had some vision characteristic.

The lamprey and the tuatara both have this same type of pineal opening and this thing is photosensitive. The structures appear to include cornea, lens and retina,

The pineal gland is weird in that it has profuse blood flow, second only to the kidney, so we can be sure that it once was of great importance. While doctors are perplexed at why this insignificant gland would need so much blood, it is obvious that whatever happened 6 thousand years ago made the extra blood flow unnecessary.

The brain of a 90 million year old bird was found with a large parietal eye and pineal gland so it's been used for some time now to provide additional insight beyond normal seeing.

Production of melatonin by the pineal gland is stimulated by darkness and inhibited by light. This melatonin stuff affects sex drive.

OK! We have a tiny organ that used to be huge and it used to be an aid in seeing, regulating moods and sex drive, but our bodies are still trying to supply it with enough blood to run a huge organ. Today the tiny little thing seems to have been abandoned by our bodies, but maybe we just can't see what it can do without vibrating a little. Vibrating allows us to understand our world.

Abraham Maslow

Anyway! This pineal gland/third eye was supposed to have given us the ability to understand the world around us. If we increase our vibrational level by unison with our environment "some call it meditation" or by other exotic means, we can sometimes get in tune with the world around us and here is the odd part. We can even affect it. Another way of saying this is that the 3rd eye thing is that "Self–Actualization" that Abraham Maslow talked about.

Positive Thinking

Somehow getting our vibrational levels in tune with the vibrational patterns of the elements around us allows us to be more intuitive. We can sense reactions needed to affect the environment. As we affect the environment we can change it. Now the changes are extremely subtle. You cannot, for instance cause money to fly off a tree, but you can somehow affect the conditions around you that will make it easier to accomplish particular tasks simply by concentrating on these tasks and believing that these things will be accomplished. I know it sounds like gobbly-gook. The problem is that the affect is demonstrated over and over and over again. Positive thinking and getting in tune with the vibrational pattern of the environment actually works. There is no doubt about it. The issue is trying to get into the level of consciousness needed to get the universe to "Bend" a little is not only hard, it also is not easily sustained once one gets to this level of consciousness.

As we raise our awareness of self, the ability to manipulate an interface between our self and the rest of the universe becomes more defined. With this knowledge we can see a "chakra level" is simply a vibrational frequency of the conscious. As the frequency increases, the amount we can affect the "life force" increases as well. This is sort of like the magnetism to electricity in the electromagnetic dynamo. While it reacts in a similar way, the "photonic dimension" and "Transfer dimension" also have a similarity in reaction to the other components of their respective dynamos. Before we can get to the transfer dimension, we still have one more chakra level we must explore and this is a wild one.

What Is Life?

Let's look at this set of dimensions more pragmatically.

Self-What people see and your carnal character, sexual interest and attraction, physical self, body, pride, hunger, vanity, smartness, the almost unnatural feeling for self-preservation, and the awareness that you alive are all parts of what we can call the SELF. These all make up a veneer that interacts with others in a physical way.

Soul-The soul, on the other hand, is the real you. That statement will take some explaining. Sometimes called the ID or subconscious, the soul reacts with and can sometimes modify how we perceive, sense and combine realities. The neat thing about this "part" of you is it doesn't die like the "SELF". It is the part that desperately love's everyone and wants to help them.

Spirit-The "spirit" can be considered as a glow. It is a window between this world and beyond. The part of us survives beyond reality. The diagram following is one I typically use to show you an initial impression of our three entities.

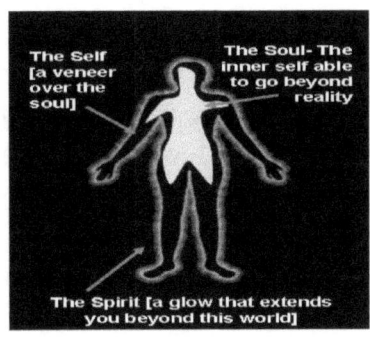

The answer to what our life dimensions are isn't simple, nor can the dimensions be defined in a simple image.

1. *The Bible says dust to dust when referring to the self or body. When our body dies, it is no more. Don't go thinking that is the end, but when you are renewed, you will be different.*

2. *The Bible says, our soul can live or die after we die. This will be described in more detail as we go along. The warning is this- "What profits a man who gains much but loses his soul?"*

3. *The Bible indicates that your soul may have several bodies before our spirit is released. Reincarnated souls are described in the Bible and many other ancient texts around the world.*

No doubt about it; the entire Christian religion is molded around the fact that the SOUL dimension of life is many times more important than the Self portion. Like Christianity, most religions identify the soul as the real entity of man. I think Taoist say it best by telling the followers that if they are to be happy, they must think of themselves [the self-entity] as an empty vessel. The Bible teaches the same thing it just says it in a slightly different way.

"If you want to live, you must die"

This isn't dying in the way we normally think of it. What the book tries to tell us is that if we try to enhance our Carnal SELF, we will lose our Soul. Losing your soul is a true death, so let's not let that happen.

Egyptians and Freud

The Christian "3 person in one" is universal in the ancient world and pretty much in today's world, so I hope none of this is too strange. The Egyptians had the <u>Baa, Kaa, and shadow</u> while Sigmund Freud had his <u>Ego, ID, and Superego</u>. The problem with all of them is the emphasis on the Self or the Baa or the Ego. The emphasis of these carnal elements is what can doom a soul to misery. I'm not talking about a lake of fire, here, I simply talking about the wonders that can be achieved with the faith of a grain of mustard-seed. Don't ignore the soul. Don't ignore the most important part of your existence. It is the part that won't die.

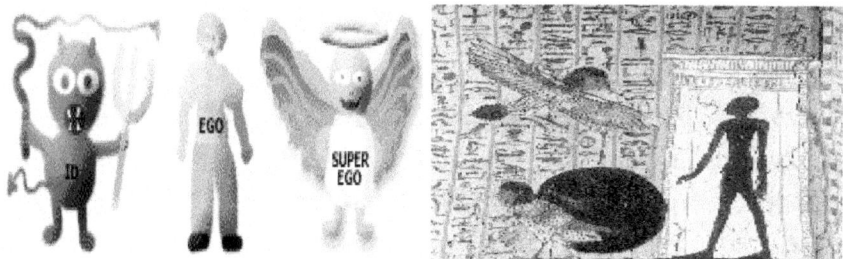

Egyptian Life Dimensions

The Egyptians essentially said the same thing as the Jews except they called the dimensions of the person the Kaa [body], *baa, sahu* and *akh* [the soul or independent attribute] , and the Shut [Shadow] the part belonging to the nether world [That would be the closest to the "spirit"]. The image above right shows how they would draw these components. The Sahu is the flying part completely free of the body. The Kaa is shown in the jar, while in this reality, it

149

is completely shut out from the "Real" existence. The shadow was, sort of, in between.

Egyptian Book of the Dead on the Ka/Body/Self- _The Osiris X, may he rest in peace, knows **the names of your ka**, the **aspect of your soul** that abides in the ground: Nourishing ka, ka of food, lordly ka, ka the ever-present helper, ka which is a pair of kas begetting more kas, healthy ka, sparkling ka, victorious ka ,ka the strong, ka that strengthens the sun each day to rise from the world of the dead, ka of shining resurrection, powerful ka, effective ka.._

The Sahu or Soul Egyptian Book of the Dead on the Sahu/Soul- "_I go round about heaven and sail in the presence of Ra, I look upon all the beings who have knowledge. Hail, Ra, I who goes round about in the sky, I say, O Osiris in truth, that I am the **Sahu of the god,** and I beseech you not to let me be driven away, nor to be cast upon the wall of blazing fire. The Osiris knows the names of your Sahu, the form in which you travel our world - the sun. Sahu pure of body, health-embodying ba, ba bright and unharmed, ba of magic, ba who causes himself to appear, male ba, ba whose warm energy encourages copulating._" [This description is exactly what I have been talking about. The soul lives beyond our body or self.] That brings us to the "Shadow".

The Papyrus of Nu -O _mighty One, when he is adored, great one among_ bas, _greatly respected_ ba _inspiring the gods with awe when he has appeared on his great throne: then may he prepare the path, justified, his ba, **and his shadow**, may they be well provided for. Let not be shut in my soul, let not be fettered my "shadow", let the way be opened for **my soul and for my "shadow"**, may it see the great god,_ [Unlike the body, the shadow was not bound to the grave

150

and could go where the body could not. In New Kingdom, tombs it was at times depicted leaving the body accompanied by the *ba*-bird. Clearly the shadow was not the soul. These concepts were slightly different that those of the nomadic Jewish people, but I think you can see the similarity. In modern times, people have struggled with the definitions because it seemed to give them less control over their environment rather than more. Sigmund Freud, for instance, tried to redefine the elements of life into his own concept to try to make it seem that this reality could hold the essence of the three dimensions of life.

Freud Life Dimensions

Sigmund Freud tried to connect the differences in characterizing a person without using ancient religion to guide him. He came close, but he missed important aspects. In Freud's model of the psyche there were the ID (instinctive unconscious), the Ego (organized, conscious), and the Superego (moralizing, not entirely unconscious) form an interactive framework which work together in the mind. Here is what he had to say. *"One of the fundamental functions of the Ego is Reality Testing – reaching into the real world to see if what is believed to be the case actually proves out – but this does not bear full fruit until the Ego has become Autonomous… substantially set free from inner conflicts between the ID and Superego."* [This is sort of backwards from all other descriptions. In his description, the EGO, or self, controlled the ID and Super-Ego that continuously were at war or mutually perpendicular.]

The ID was the evil characterization while the Superego was close to the definition we must place on the Spirit portion of the body; sort of the HOLY component of a person. Unfortunately, he got the ID and EGO backwards. If any

151

portion of our self would be considered evil, it would be the Self portion or our conscious mind. Of course the ID or self is not specifically evil, whatever evil is. It is simply carnal. With that as a background, let's get into a new description of our universe so you will see how everything has a place and everything is required for our universe to operate.

Conservation of Life Force

From these we interpret the continuation of the Soul and spirit after what we call death. Life force is conserved just like Energy and Time.

Conservation of Energy-I think I had better go over this again. According to Einstein, vibrational nothingness associated with mass and light continued outward from a central point and escaped the limits of the universe at infinity. This means that sooner or later, all will be dark and no mass or energy will exist. While that would certainly be in the distant future, it still makes me uncomfortable and it really doesn't make sense that existence is not renewing. Additionally, conservation of Energy laws would be violated. As the light energy hits the end of our universe and leaves, it is immediately replenished. Everything else in the universe renews and we can, pretty much, be assured that the same is true of mass and light [electro-magnetic stress]. Everything in this universe is continually renewed.

*If energy is renewed, *life must be renewed* in a similar way so we can gain information about ourselves by looking at non-living energies as well.*

Conservation of Time-Time must also be rejuvenative. Here's the deal. Time must go backward while it is going forward. I know that sounds weird. This will be an interesting concept when we look at death in particular.

152

Does death come first and a person gets younger or is it the other way around? Are there just snapshots in time and only the now is real? If the now isn't real, are our "selfs" real? Maybe Chromosomes can help.

Dead Chromosomes

A dead chromosome and a live one look the same and have the same characteristics. The sugars are the same, the links and bonds are the same.

Something outside the chromosome makes it alive.

This thing outside the DNA and Chromosomes is what we can call life. In particular, it is the "self" driven by the "soul". Both seem to be "Conserved" in this universe. As one set of chromosomes or a person becomes dead, another seems to become alive. I know you are thinking there are more people alive today than during the old days, so this "Conservation of Life" doesn't look like it will hold water. This is because you are not identifying life without self. Just because a soul has no body does not mean there is no life in it. It simply cannot experience what we call reality just like gravity can never become matter. We can think of this state as going the speed of light. At that "speed" all life is suspended; you do not age at all as this reality is completely gone.

When one talks about people vibrating, it is both simple and almost impossible to understand. Everything---I mean everything, is made of vibration. This includes all matter, all electro-magnetics, all nuclear energy, all photons, even all life forces and those we would consider dead... That by itself is not enough information. The second thing is that live and "dead people" control what we believe to be reality. Let me make a simple observation. ---The color RED--what is it?

The answer, of course is vibration and more specifically, electromagnetic vibration just wiggling all over the place. While it is wiggling at a certain rate, other colors and things that are completely invisible are vibrating at different frequencies in a constant time observation. Why in the world do we see RED when it is only vibrating nothingness??? The answer has to do with that Anthropic Universe things I mentioned before. When I say anthropic, I don't mean the "seems to" definition people sometimes attach which completely destroys the truer meaning.

Participatory Anthropic means that our linked consciousnesses define observations to vibrations. [Notice I did not say "seems to". In fact, our group consciousnesses invent ALL of the [carnal] reality. Carnal meaning the reality we see, smell, hear, feel, etc.

Reality Without Coordinated Resonance Of SOULS

This "invention of the souls" is what we call reality!!! Imagine a reality where all you feel is vibration, there is no light as it is just vibrations and there is no distance to allow you to know where you are. If you try to interact with another SELF, their perceptions of vibrations are totally different and you don't even know the 2 of you are nearby. What good is the vibrational cluster of mass or the vibrations of time all around without the "stage" we call reality?

154

Reality

If you have ever heard the terms "Power of Positive Thinking", "Think and Grow Rich", and all other concepts of the 70s which tried to convince you that how you consciously view reality will affect reality, are not only true, they affect your death as well. In the Anthropic World if you have faith of a grain of mustard-seed, you can move a mountain, as Jesus said thousands of years ago and you can walk on water as demonstrated by Elijah, Elisha, Peter, and Jesus so many years ago. With the Anthropic Principle, science and religion can act as a single tool for us to understand God, the universe, and ourselves. The dead are not the only "souls" building our reality. Live people "souls" just like dead souls shape and mold reality. Let me give you an example.

Eating

Today we know that what something looks like changes greatly what something tastes like. If you see mashed stuff and it tastes crispy, the "mind" quickly determines what it SHOULD BE and viola' it ACTUALLY becomes "that" to the taste buds, touch sensors, and emotion centers determining that you like of don't like what you are eating and level of satisfaction you feel after feeling the substance. The taste bud only looks for a chemical so it can vibrate differently. One taste bud wiggles from salt, another from a sour, etc. nothing really determines taste beyond these minor elements. The wiggling is in the form of chemical combinations allowed by structure. If one of the "particles"

155

can attach to the crystal, it gets bigger and vibrates fasters making an electrical differential which is "felt" back to nerve centers in the brain and it is magically determined if it was a good or bad taste.

Have you wondered how in the world some people "love" the taste of nasty stuff and you can't stand it? If there was a reality to taste, all would either like the brain response or not. Taste is simply not real.

The same can be said for seeing colors, sensing Aether, looking at beauty, etc. It is all pretty much fake and only "defined" by how we interacted with this "reality" over the years.

Being

What is the perception of being? One answer is seeing the things around us and interacting with them. A problem is that there are no real things around us but simply vibrating nothingnesses as Einstein and others have proven. The forces established by the vibrations give us the "perception" of mass. One can assume that even after the soul is released; it could conjure images of reality, especially when there are groups of these souls that are together. Later we will examine souls between lives if they are not in a state of sleep. Right now let's continue that live reality.

Electricity and Sight

For this discussion, let's look at sight. Vibrating of something we defined as electro-magnetivity makes a chemical change in our rods and cones of the eye. The chemical changes produce something we call electricity which excites portions of the brain. Everyone uses this electricity stuff to define everything, but it doesn't exist. It is a "potential to do something" by definition. It does not exist except for something we call work that requires some outside intervention with this "invisible potential". It is only the magnetic field produced as the electricity changes that makes it real to us. While it is changing or "VIBRATING" it is in our perceived world. The brain remembers what it perceives from the changing electrical signal and simply "defines" what we call sight. Just think what our world would be like if we could REALLY see what enters our eyes. The things we think we see are just reflections of some external light source rather than the actual object. When the external light source is removed, the objects we see are removed from view. Try to remember that the "real" reality comes from a thing you can't see called the SOUL. The entire universe is controlled by the joining of souls all defining away.

Our existence or what we perceive as our existence is a combined implication of all in existence. This includes existence when we are DEAD.

Faith and Reality

I know I have laid out this anthropic world and told you generally about how the universe is modified by people and you have totally believed what I have been saying because it makes so much sense. You are skeptical. At best, you looked up a sight on Anthropic Universe and found out that one way of looking at it is to sense that the universe was simply made for us. God grabbed the universe and modified it so that people could be created. All the creationists roared, but that is simply not the end and if you try to define people that way you simply have something like mashed potatoes oozing around to fill limitations of the universe.

What anthropic physics really shows is that people can modify the universe [to an extent].

Over time, the universe, quite naturally is shifted to be in line with the needs of people. I'm going to tell you how this also affects what you call death, but you first need to broaden you awareness so the details will be more useful to you.

Faith is not Faith

Remember, Jesus told his disciples that with faith of a grain of mustard seed one could move mountains. Here is what he did not say. "Faith in me will get me to move mountains." Surely, he could do that whether we had faith or not. If he wasn't talking about faith in him, what was he talking about? That, my friends, hopefully, is becoming more evident as we go along. Buddhist monks, for instant, have substantial amounts of faith, but they have no regard for Jesus, God incarnate. These monks have done miraculous things. The Gurus seem to have something we could identify as faith, but they also have no specific faith in Jesus. Many of the faith healers around the world don't profess any specific religious order. The Egyptian magicians Jannes and Jambres had no faith in God, but they could turn a stick into a snake. On and on we could go. Faith, as discussed by Jesus in his plea to his followers was something besides faith in Jesus. I am certainly not saying don't have faith in the living God and God Incarnate. That is a different subject. This is simply saying Faith allows us to change what we might call Space Resonance.

Resonance

I have been talking about resonance throughout this book, but let's take one more look as resonance of all dimensions together is what we call reality. I think I have you worried right now with the first description of how matter and electricity are correlated. Let me back up a little and restate resonance for this application in the words of Dr. Milo Wolff who is one of the leading master physicists who has greatly extended Einstein's initial work into a "usable" platform. My comments are in "bold".

"Resonance is composed of a spherical IN-wave which converges to the center **[of the universe and comes from a different universe as a component of the operational dimension dynamo]** *and an OUT-wave which diverges from the center* **[of the universe and makes up what I call the structural dimensional dynamo]**. *Their separate amplitudes are* **[close to]** *infinite at the centers.* **[Like all other resonance factors in the universe, how close they are to being infinite can be considered the "quality of resonance".]** *When combined, the two waves form a standing-wave which has a finite amplitude at the center. The standing wave* **[appears]** *to be the structure of the electron. The inward and outward waves* **[sort-of]** *provide communication with other matter of the universe. Spin of the electron is a result of the reversal of the IN wave at the center to become the OUT wave."*

While there are still limitations, this, this definition helps us interpret how an adjacent universe "establishes" resonance in this world. The more we communicate with an adjacent universe the faster our vibrational resonance becomes and its quality rises.

Quality of Resonance

Let me explain this "quality of resonance" a little because it is this Quality that will allow one to expand how a soul interacts with reality. In electro-magnetics, quality of resonance describes the difference between the effect of a circuit outside its resonance frequency and that which can be described when it is in resonance. If a crystal is excited with a vibration that is half of the frequency it likes, it may vibrate a little and nothing more, but if it is hit with the vibration it likes, it begins to self-oscillate substantially. Just think of a tuning fork and how it always sounds the same when struck. That is its resonance and the longer it makes the sound describes its "quality" of resonance. In the electro-magnetic world, this "quality of resonance" depends on many things including what the crystal is attached to, how well the crystal is cut and how homogeneous the crystal is. In the electron or particle world, the same things can be surmised. Purity of the particle and the things that surround the particle affect how close to infinity the standing wave appears.

Resonance and Matter

I guess you are wondering why I even brought up this resonance in the first place, but **resonance holds matter together, it holds time together, and it holds life together**. If enough electrons are in an area that are sensing similar in-waves, they align together just like a crystal. One could say that atoms are resonant plugs that are held together by "like vibrations". Scientists found these things called <u>gluons</u> which seem to act in opposition to other particles and quasi-particles. Gluons hold quarks together. Three quarks and an

161

unknown number of gluons are called an electron. **Gluons are quasi-particles [fermions] that have a negative gravity.** That is, the farther the quarks move away from the gluons the STRONGER the gluon attraction becomes. It is sort of, like the quarks are inside and invisible piece of matter that has gravity. The closer they get to the surface of this invisible piece of matter, the more the gravity affects the quarks. The center of what we call a gluon would be the resonance point of what an electron whose resonance is defined by the vibrational characteristics of its component parts. I know that was a mouthful so let me state it differently.

Gluons are not odd, they are simply invisible. One can say that they are an in-wave and out-wave collision; sort of the core of an atom.

Life Resonance

There is a reason I keep bringing up religious testimony. I want you to see that True science and religion both ARE the same and I'm going to let you in on a secret. Consciousness/ life/ and death all act the same. If we wish to affect the universe more and have a higher quality of resonance, we must become pure and surround ourselves with things that allow this pureness. OK! I can't exactly define what pureness is, but prayer and meditation is probably more important to our quality of resonance and our capability of affecting the universe than one would initially believe.

Who Cares About Resonance?

Why have I even brought up resonance? If a vibration node gets larger or smaller shouldn't matter to us. Right?

Well----we need to care for a number of reasons. Here are a few.

162

1. **The higher the resonance** of electro-magnetics, the closer it comes to being pure magnetism. Even at the somewhat lower levels of resonance, depicting light is considered the **most useful** form of electro-magnetics to the universe.

2. **The higher the level of resonance** of a particle or quasi-particle, the **more useful** the matter it becomes. The higher vibrations mean larger and larger particle including god, uranium and the powerful nuclear materials. Eventually matter become total gravity with no real nature associated with mass.

3. **The higher a person's resonance is**, the more he or she **can affect the universe** and their own characteristic universe becomes more in sync with everything else. The closer one can get to God. Buddhists indicate this change in resonance of a person's "life" as chakra with the highest level allowing someone to completely separate from this carnal world.

Meditation and concern for others rather than yourself raise the resonance of your "Life" which puts you more in tune with the "spiritual" reality. That is a good thing.

Time resonance could also affect us, but it is the differential where some move faster than others. Vibrational frequencies discussed here actually do affect time resonance, but it cannot be noticed from inside the time frame.

Resonance & Quantum Mechanics

Niels Bohr [1885-1962] tried his best to mess up our minds. He indicated that there is random vibration associated with everything, but if you test the spin of an object, and it is going one direction, there must be another that is indelibly linked and going the opposite direction. If that wasn't odd enough, he indicated that the linked objects could be millions of miles apart when the transposition occurs. Einstein indicated that the whole quantum mechanics theory was junk science that violated all of his theories. Einstein died before tests were devised that generally proved the phenomenon. --- Not only had the linking be seen, but also this elimination of time space as a boundary has been witnessed.

Today, scientists have witnessed sub-particle teleportation that eliminated time-space.

Without "space", vibration has a difficult time being defined. Its definition needs the introduction of a space based REALITY. Before he died, Niels was writing something entitled "Light and Life Revisited" what he had finished was published after his death, he understood how very close light and life truly are and this quantum mechanics thing may hold a key to help us understand it as well. My version of light and life is a little different than Niels, and it has a lot less math, but the idea is similar.

Eliminate Time and Space

I know you are saying--- "What in the world has he been saying?", but I think some of it will begin to make sense as you go through this book and miraculously, you will understand Death better. You may also be saying, "What in the world causes the elimination of time and space in quantum mechanics?", so here is one theory. In-waves that introduce force and out-waves that establish physical characteristics are each associated with "times" that are backward to each other. While this allows work to ensue, it does something very important besides that. **It eliminates time during the exchange.**

Simply put-- backward and forward time together make the absence of time and quantum mechanics can be more easily understood as these two "times" come together. Putting it another way; quantum mechanics requires a second universe to neutralize time.

Because there is no time generated during the exchange, the exchange cannot be recorded by us. Even if the exchange took a thousand years, there would be no recognizable time for the action. Here is the scary part. Even if one were to go the speed of light there would be no recognizable time.

What?????!!

[Sorry for that last question mark and the 2 exclamation points, I was getting frustrated.] Here is the leap of faith. If there is NO TIME, there is no space. The actions would be resident at any or all locations at once. This, I believe, is where Einstein began nodding off. He was old and tired and unwilling to even consider the elimination of space. Neils Bohr was still young and thought about all of this weird stuff.

165

What this does is--- allow us to transfer- to anywhere-anything----at least the essence of anything--- instantly just like the Star-trek hyper drive.

The reason one could travel the distance is that there would be no distance so long as one could somehow introduce in- and out-waves together. That is where resonance comes in.

If someone wants to introduce a modification of our "reality", all that is required is to temporarily change ones resonance.

Resonance in the Old Days

Jesus, his apostles, Moses, Elijah, and others from the Bible would have used this simple technique to turn water into wine or blood. Simply change one's resonance and something has to give to correct the shift. Do it right and things seem to be miraculous. Don't get me wrong about what I'm saying. As I stated before, Jesus **was and is God incarnate**, but while he was on Earth he was 100% human and he taught some to understand what he meant when he said "With faith of a grain of mustard seed **anyone** could move mountains."

He could just as easily have said that changing your resonance by elevating your awareness to the non-carnal universal of heaven would allow one to control many carnal things simply by removing time-space during the transition for this equalization known as quantum.

I guess his apostles looked at him weird enough when the moving mountains comment was made, so he tried to keep it simple.

Vibrations of Existence

We know that the faster the photon of light thing vibrates the more powerful it becomes. Soon the fast vibrating photon thing becomes dangerous to humans as it can go right through the body [x-ray] and if it slows down too much it changes into something we call radio waves. I know you are thinking that these radio waves must not exist because they don't produce light and they have no mass, but let me assure you that sometimes these photon things do act like normal matter. If you look at the diagram following, there is a wiggly line. The faster wiggling represents a prime particle vibrating faster and faster. Radio waves turn into light that turns into the deadly gamma rays. Vibrations of the Soul dimension do the same thing, but it changes the characteristic of our perceived "reality".

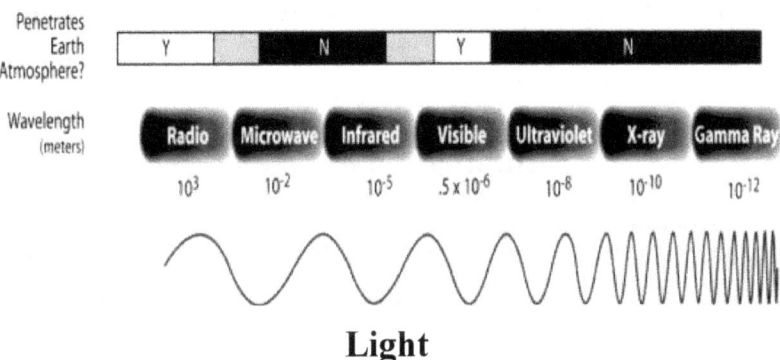

Light

Today, the chart can continue even farther in both directions as light starts its vibrational journey as static electricity. As vibrations start to increase, we call the outcome electro-magnetism. When we get to the highest vibration, we can define the vibration or thing as pure <u>Kinetic Magnetism</u>. All three dynamos have similar static and kinetic limits and dualities with the other 2 dimensional dynamos.

Life and Matter are Similar

As the ethereal dynamo goes through the same type of transformations from the **static carnal life** to the **kinetic soul life** and matter goes through this same transformation from the static Aether up through the pure kinetic gravity of what we call a black hole. We must get a sense that the 10-dimensional universe concept seems to pull everything together.

Resonance of light and Life

It is reasonable to assume that ultrahigh frequency [Kinetic] light, black holes and the soul are joined by vibrational similarity as are the complete carnal life of a tree; the Aether that could eventually make matter and the potential electric fields [Static] that could eventually do work.

For light to provide a resonance and be sustained the surrounding vibrational characteristics of the other dimensions must be similar and this is how our soul can modify reality.

Elimination of Gloom

Let's go slowly. One way this might be attributed is that when we sense the light as "**warm or comforting**". This is because mass, Force and [your life] are in tune. If you bring in gloom, the same sights will become "Gloomy". In fact, reality around you will be more "depressed". The "power of positive thinking" isn't magic, it is resonance. As you vibrate your soul to a higher, [less selfish, less, carnal way] you will be happier and reality around you WILL be brighter. When a majority of the life-forms are resonating at a similar vibration level, they are unified with a perceived reality and each can interact with the other. Becoming too debased and selfish, or too holy changes everything. Those on the debased side are pulled along barely changing reality in any way. Abraham Maslow looked at the other end and found that people who come out of the gloom can become "Self-actualized". By not thinking of self, the self becomes more in control of the environment.

If we can vibrate faster, awareness of a higher level of light can bring comfort, insight and understanding. This sounds like all that "power of positive thinking" stuff and it does have some similarity, except that vibrational base is what we call life instead of light. Life and light are very similar if you have not noticed. What I mean by this is that visible light is a moving thing. It goes beyond sight, and extends into comfort, desire, understanding, and awareness. Light even can explain matter because all dimensional dynamos are intricately locked together. As one dimension is stressed outside its normal resonance, the others MUST follow suit. Hopefully this is allowing the whole Anthropics things sound more feasible as there is a substantial amount of evidence for the controlling element of our consciousnesses building this universe and modifying it.

Particle Resonance

I need to drive this stuff home a little and to do that, let's look at particles.

The difference between a helium atom and a gold atom is vibration, but what keeps the gold atom together? The answer is resonance.

Just like the electro-magnetic resonance, particles express this same feature. Particle resonance is the most comfortable frequency for the Aether [potential for having mass] and gravitational field to stay. It is a point where Aether and gravitational fields both have the same strength and when that occurs, the effect of the 2 fields is most stable. If these two fields are stable at a high frequency, they appear to be a large atomic mass. At lower frequency resonances, the lighter atoms become apparent. You might have gathered from the similarities of the various examples that the esoteric components of life and consciousness would also have this resonance feature and its manipulation can be described and so it does. In this case, a life force and consciousness level would be matched to provide the most stable life pattern.

At the Speed of Light No One Ages

Let me try something a little different in describing life resonance by using light. While light is constant, we have also determined that there is no aging if someone is going the speed of light to us. It makes no sense to us, but we sort of accept it. Atomic clocks being sent in space ships come home showing the "experienced time" in the space ship was reduced. From these experiments and others, there is little doubt that life is, somehow, suspended at high velocities. Unfortunately, it is not a simple observation. If someone was going to the nearest star at very close to the speed of light, he would get there in about 4.5 years. We could watch the event and record the event, but the man in the ship would still not age. His universe slows down as he speeds up; to him almost no time passes. If he shines a flashlight about halfway to his destination, the light from the flashlight would get to the destination before he got there in his world, but about the same time for all other viewers. This reduction in aging is the best evidence that the Life dimension is associated with individual universes or associated with universes linked by a common velocity. Remember, it doesn't matter what direction you move to cause this affect. If you were simply **vibrating** at the close to the speed of light, you would age very slowly and everything around you would simply start rotting before your eyes.

Carnal and Spiritual Life

All that aging stuff is carnal living, but some aspects of living are not carnal, they are spiritual, and sometimes the spiritual aspects of life become very pronounced. They don't follow what we normally believe to be truth, but that does not mean they are not real. People really can bring others back to life, turn water into wine, walk on water, and lift up cars to protect those they love.

What Is A Life In-Wave?

I've been kicking around these vibration "standing waves/nodes" and these odd in and out waves, because some physicist identified them that way.

I'm sorry!!

Let's try to put a perspective on them. The out-waves are fairly easy to understand as vibrating nothings emanating from each of these standing wave point. OK! Not very easy, but at least one can generally understand the concept. Life out-waves are generated by conscious understanding of the environment. [Self, Survival, and Sex]. None of these attributes allow modification of the self-resonance and all can be identified as CARNAL or generated from inside the universe creating out-waves that spread out to the limits of the universe. [The body is inconsequential]. OK it's nice to have a body, but what I mean is that the Self, Survival, Sex life of a bug or germ or even a person doesn't affect the other dynamos. They are Static and generally identified as "the self-dimension". The out-waves keep going away from the "node" to infinity and --------beyond [never to return???????]. It would seem that life would vanish from the universe over time.

Life Doesn't Vanish

For life, out-waves constitute the carnal aberration of living. All aspects of self can be easily shown to have initiated in the universe and as they try to expand outward to our adjacent universe, then something magical happens. They are bombarded with in-waves of life just like particles were bombarded to establish stress in particles so that work can be done in the universe. This bombardment alone should allow us to know that other beings are resident in the adjacent universe. Their survival dimensional vibrations continuously leave and are converted to in-waves that CAN react with ours if we want them to.

Some More from Milo & Albert

To try and get a feeling about what in-waves are, Einstein and Dr. Milo Wolff will help. Dr. Wolff stated," *Forces/Fields are caused by wave interactions of the Spherical In and Out Waves with other matter in the universe which change the location of the Wave-Center and which we 'see' as a 'force accelerating a particle.'*"

In English, this means that these in-waves make force to establish energy needed to run our universe. Unfortunately, there are a couple of real issues. Where do these all important in-waves come from and how do in-waves relate to living?????

172

These In-Waves Are Different

As Carnal life spews out of this universe, we can assume it can be a driving force in our linked universe that has living substances spewing out Spiritual life that regenerates our Carnal living.

Life Out-waves

You may know what I'm going to say, but I will say it anyway. The in-waves are coming from an adjacent universe. When they come into our universe, they act backward to the out-waves produced as various out-waves come in contact with each other to produce vibrational nodes we describe as atoms or atomic clouds. <u>Because the in-waves are backwards, contact with "out-waves" cause STRESS we call conscience.</u> Think of it this way. Carnal out-waves stressing self/survival/sex emanate outward and they are countered by in-waves that stress the exact opposite which are true love, true concern or empathy, True willingness to die for another, true spiritual insight. The next time you hear a little voice telling you not to be selfish, or eat some food you should not have, or giving you some super strength to save a life, understand what that really is. It is simply the In-waves of life just like we have been going over and over, as they are represented in the other dynamos. One can think of them as backward selfishness, or backward hate.

The in-waves are not backward to the adjacent universe. Time is reversed to assure conservation of time and life is reversed in the same way. In the adjacent world, the Self dimension IS the Spiritual awareness/empathy, and love. The out-waves from our universe are mostly self/sex/and survival which are used by them to allow them to understand themselves in some way. If that isn't odd enough, as time goes forward here, the concept of time in our neighbor MUST BE backwards to us but forward to them. Because the in-waves are in opposition to all the in-waves, when they come in contact, they put stresses on the life's vibrational nodes. That is what we think of as <u>carnal stress</u>. Carnal stress has limited effect on our lives unless we can change our life resonance. That is coming up but it may start sounding religious so I'm just warning you now.

Life Would Disappear

If we had no life energy inserted from out adjacent universe, life would simply have no meaning at all.

If someone tried to inform you of one basic truth about life, it would be that living without a heaven is simply existing and life in our universe would eventually disappear.

So what! You created a great food dish or ran a mile, or saw a sunset, or became an evil dictator. The motions of life would simply have no meaning, they would be actions controlled simply by the environment. The word predisposition would be will rather than could be. There would be no moral or immoral. There would be no good or bad. There would be no "substantial" happiness or sadness. I say substantial here because humans and robots, for that matter, can fool themselves into thinking they were happy or sad or depressed or destitute. Possibly even a germ thinks he is having fun.

In and Out Wave Differences

The reason out and in ways are opposite is that adjoining universes have "backward time" to each other. I know I've stated this a couple of times, but it needs to be understood. As an out-wave leave our universe, it appears to reverse by

174

the time dimension of the outside universe and it is turned into an in-wave. Guess what!! The out-waves in that universe have vibrations in opposition with the newly created in-waves. As these out-waves leave the adjacent universe they enter our universe and appear to be opposite because of our backward time to the other universe. Of course what that means is particles in this universe become forces in the adjacent universe.

When studying life, one must also sense the difference in experience. Out-wave "carnal" experience is substantially different than "Spiritual [in-wave] experience. Both must be levied together to allow for a more meaningful life experience.

God knew that Carnal existence and the feelings of self, sex, and survival were too strong so one of the three dimensional components of GOD was sort of introduced to channel the spiritual in-waves into our carnal selves.

- There is always a struggle in ourselves to do the right thing followed by doing the most comfortable thing.

- There is an innate carnalness to mankind, no matter what we say.

- We know that consciousness does affect reality. It has been somewhat proven in experimentation and substantially in calculation.

- If life did not have a way to be rejuvenated, soon there would be no life in the universe just like there would be not light and all matter would be in a state of entropy.

- The power of positive thinking and self-actualization absolutely do change our environment just as written about in dozens of books.

- Life is more than chromosome multiples in a body. That is simply stupid.

- Life is more than chemical reactions in a brain. Just like sight and vision are completely different, life and predispositions are completely different.

- The will to live is a very strong force that has nothing to do with the brain.

- The conflict of one's thoughts to do good and/or something that will satisfy self is not caused by chemical reactions in the brain.

- When your body dies, life does not. It either stays in this universe or it goes to the adjacent universe.

- Some methods were discussed in the preceding book to allow travel into the adjacent universe to provide a mode for time travel.

Crown Consciousness

According to Buddhists the highest vibrational Chakra is the Crown. As people expand their awareness more and more or become more self-actualized, as Maslow said, we become more and more conscious of other worlds or universes. Our conscious becomes more in tune with other universes because they are an integral part of this universe Guess how people describe how to change to higher and higher chakras. They indicate that each expansion is like a vibrational realization. The more the realization or level of chakra becomes; the more noticeable is this strange vibration. The method for getting to this "crown" thing is to sort of hypnotize yourself. "Relax! You are getting sleepy! You are completely unaware of your surroundings! Your feel a warmness and a sense of comfort! Wait just a minute! I'm getting numb here and I can't do the crown chakra dance right now. I have to finish this book.

We are told if you get near this state, the body will begin to hum and vibrate in a low comforting tone and it almost warms your entire body and your feet and hands seem to disappear. Wow! The first time it happens, the sense of comfort is great but nothing more may have happened. A second and third time might be tried and the world around you is forgotten as you slide into the vibrational level associated with the crown chakra. Vibration, vibration, vibration- if that isn't the weirdest thing I ever heard. Hold on just one minute, vibrations are not odd in this book. Vibration frequencies define how a person perceives this universe just like the vibrating fermion that begins to perceive the other particles around it. Let's think of this whole consciousness a little. Let me start over with a question.

Can your consciousness REALLY leave your body?

I'm sure your first thought is that it can't but don't be so ready to close your mind to things that seem to be going on around us. It is becoming more and more apparent each year that astral projection, near death experiences, seers getting their prophesies, and even reincarnations have been and are elements of the same characterization of the consciousness dimension. I know you think you are using your consciousness right now, but there is more to it that you would like to believe. Let me give you a few examples.

Near Death Experiences

It is believed that over 10 million Americans have had Near Death Experiences and lived to tell about it.

One need only go to the near death experience website and find 2000 verified or at least printed events involving near death experiences both good and bad. The website is [http://www.nderf.org] for those interested in a first glimpse. Below is a common theme in just about all of these things.

- It is said that the soul [I call it consciousness, but some like to separate us a little more.] goes through this tunnel like "Whoosh".
- After the whoosh feeling you are standing in the brightest white light you have ever known. The noonday sun cannot compare to its brightness or stark whiteness.

- You instantly feel this bright white light raining down upon your spirit
- You feel an intense love all over your body like soft rain falling on your skin. You know you are loved beyond all shadow of a doubt by this bright whiteness surrounding your spirit.
- You feel totally at peace and very safe and love is in you and around everywhere.
- You would be typically calm and have no real thought of whatever had "almost" killed you, no pain just absolute peace.
- It is said that you feel odd about still thinking, and how alert you are.
- Someone may ask, "Do you want to stay or do you want to go back?"

Out Of Body Experiences

If we were just looking at out-of-body experiences in general we would find that they are much more common that we would initially believe. This thing occurs in about 1/4 to 1/3 of the population depending on which study you look at. It would be ludicrous to say that up to one third of the human population are mental illness deviants, when in fact, this is such a common phenomenon. They leave their bodies. They see things at great distances from where their body is. They talk to "friendly and informative" people. They insist that the time they are away is not dreamlike, but instead it is close to reality. They usually sense power and freedom. These people recognize and describe objects seen in these states with great accuracy. Others, including many who initially were very skeptical, have verified this strange fact.

Astral Projection-One type of out-of-body experience is call astral projection. Below is a common projection memory.

- Over several days a possible projector may try to focus on some special place.
- Many concentrate on a mantra [some special relaxation word, phrase, or image] prior when falling asleep.
- The mantra is used to sort of allow fast self-hypnosis or allow the body to fall asleep faster.
- A mantra is said to also help a person stay conscious enough while in a dream state to have control to some level.
- The person would feel their physical body fall asleep while the image of an astral body would emerge and start to rise.
- The astral body could go up and out of the experimenter's body.
- This is usually easier when the goal was clear.
- One can usually feel themselves flying through the air.
- As with other out-of-body experiences, sometimes other people can be seen or even talked to during one's travels.
- It is said that caution must always be used not to fall into deep dream state at this point.
- Many not only view remote sites, but they feel like they gained some personal teaching that stays them in a strong way after the experience.
- They, generally, feel at peace when they return to their body.

Common Thread

Hopefully, you are seeing that in almost all cases, people begin these experiences by blocking out the world including all feeling. Those who are forced in that condition by some tragedy don't seem to have any difference in this the effect. They leave their bodies, get comfort or wisdom, sort of talk to comforting people, get a heightened sense of reality, can float, and when they get focused back on the "real world" they are plummeted back into it. Many times these people are changed forever. The trip to the Crown chakra has changed them forever and I'll tell you why. Their consciousness has been vibrationally enhanced. It vibrates closer to the level needed to do this transfer thing, but in the mean time they become more aware of the feelings of others and become more self-actualized. Whether the "people they interact with are the cause of the vibrational enhancement or some other mechanism is at work, I do not know, but the entire life force of the person is enhanced. Many like it so much they go off and do it again if they can. Prophets and seers seem to do it a little differently.

Seers & Prophets

Something that is similar to astral projection seems to be the ability to tell the future. For that discussion I will bring out some details of Dr. Edgar Casey, and others. While it seems that the seers do leave their body, there is a substantial shift in what happens. Like other out-of-body experiences the seer must generally forget the details of this world. Some of these guys got to the "Crown Consciousness" by going on long fastings. Nostradamus stared into a pool of water for long periods of time. The more they separated themselves, the faster and more accurate were the prophecies. While in this altered state, the seers are somehow given the information. They actually witness the events that would happen in our future. There is one likely reason that Nostradamus, Mother Shipton, John and Daniel from the Bible and many others all saw and described the future. That reason is BECAUSE the future had already happened at the location that had "projected" themselves to.

Dr. Casey's Time Travel

Dr. Edgar Casey simply fell asleep to get his answers. While he was in his "trance" he could answer just about any question from medical miracles to future events. For years he was able to amaze, comfort, heal, instruct and help others simply by leaving his body for a while and getting answers from "somewhere else. I don't know if he feel though a tunnel and came to a bright light or any of that, but he did see the future. It is obvious he didn't understand some of the

182

things he saw, but there is enough details in his ramblings that let us know he truly saw the future well before it happened in this universe. Some of his predictions have come true and some still wait as indicated below

The Stock Market Crash of 1929 was foretold in February 1925.

In 1935, Edgar Cayce warned of World War II calling the Germans the Aryan race.

The Beginning of the Earth's Poles shifting was predicted in 1936. He actually indicated that the shift would begin around 2000 to 2001AD. Sure enough we now know the shifting has begun.

Convergence of Communications Companies was predicted in 1929 and seen "Ma Bell began its quest about 5 years following the prediction.

The Dead Sea Scrolls were predicted as Edgar Cayce mentioned the Essense 10 years before the Dead Sea Scrolls were discovered in 1947.

Blood was predicted to become a Diagnostic Tool. He knew it would happen in 1927.

La Niña and El Niño Effects were explained. Amazingly, on May 28, 1926, Edgar Cayce connected the temperature changes in deep ocean currents to weather changes in the United States. I know it sounds ridiculous, but that is what he predicted.

Some of his predictions have not happened yet, but things seem to be looking more like the descriptions of this unusual guy.

In 1926, he predicted that our life-spans would extend. I know we are still struggling, but with all the body parts being cloned and gene splicing, the ravages of time will soon be lessened. None of that was happening in 1926.

He predicted that we will discover the design for a self-fueling perpetual-motion machine. Many designs are almost there today.

In 1941 he predicted a major conflict in the Persian Gulf. He actually indicated that the areas to look for would be Libya, Egypt, Ankara, in Syria, and in the Persian Gulf."

We will gain expanding consciousness was predicted in 1942, He said that individual had to essentially work with personal soul development in order to ***resonate*** to that new consciousness: This gives us insight into the method for gaining insight. He would somehow allow his body or soul to vibrate to a higher level just like many are now trying to indicate [Besides, it goes along with my book.].

In 1932 he indicated that an ancient "Hall of Records" would be discovered in Egypt. He said that the people of Atlantis became aware of the fact that their civilization was about to be destroyed. As a result, they hid identical records of the Atlantean civilization in Bimini, in Egypt and in the Yucatan. The records contain a record of Atlantis from the beginnings of those periods when the Spirit took form, the first destruction and the changes that took place in the land. He even identified exactly where. He said that it is located where the sun rises from the waters, the line of the shadow falls between the paws of the Sphinx.

He told us about China become a world power. He said that eventually China would become "the cradle of Christianity, as applied in the lives of men and that it would take a long time to manifest but that it was the country's destiny.

He prophesied about the Second Coming of God. He not only told us a lot about the "missing" years of Jesus' He indicated that the time and half time has been fulfilled in this day the Lord, will come, even as they had have seen him go. I know he could have looked in "Revelation", but it did give us an indication of his Christian faith.

Mother Shipton's Time Travel

Let me reintroduce Mother Shipton for those that don't recognize the name. Her real name was Ursula Sontheil. Her last name was hard to pronounce so she married a guy named Toby Shipton. Now Ursula was not pretty at all. Warts and all the witch-like looks were indicated in documents and drawings so people believed she put a spell on old Toby. After he died, she became more recluse than she normally was and she started living in the future so to speak. She wrote cute poems around her predictions, but they are interesting in their own rite. We don't have details about how she vibrated into a new level of consciousness, but we do know some of the things she saw. Let me share one of Mother Shipton's visions about a new queen that would come to the throne only 3 years before Ursula was burned to death. You will notice Queen Elizabeth I, Francis Drake, The defeat of the Spanish Armada and other things she simply could not have known about as they occurred after she had died.

The Maiden Queen full many a year- Shall England's warlike scepter bear. Those who sighed, then shall sing.-And the bells shall changes ring. The Papal Power shall bear no sway. & Rome's trash shall hence be swept away. The locusts from the 7 hills This English Rose shall seek to kill- & the Western monarch's wooden horses -shall be destroyed by Drake's forces.

Besides her own awful death, she predicted the following:

- *The rise of the Church of England-*
- *The California gold rush-*
- *Automobiles,* *Radios,* *telephones,* *telegraphs, hydroelectric power-*
- *Manufacture of mountain tunnels* *and Commercial air travel-*
- *Submarines,* *airplanes, iron ships, & airborne military and their use-*
- *World War I,* *US Civil War, and the French Revolution-*
- *British and French alliance* *during World War I and II-*
- *The Allies and Communist* *bloc, and the cold War-*
- *The France to England* *underwater tunnel-*
- *Women would commonly wear pants* *and have short hair, [an unthinkable thing at the time. OK Ursula thought it, but she still wore the huge pile of clothes that every other woman wore.]*
- *The printing press* *and how it would change writing*
- *She saw the coming of a comet* *that will sort-of begin the spiral downhill for mankind.*
- *After the comet Mother Shipton saw war,* *famine, tyrannical rule, the tribulation period and finally a long lasting peace just before the eventual end.* [This was very similar to one of her contemporaries named Nostradamus and the marvelous revelation of John in the Bible as well as some of the details in the predictions made by George Washington when he was given his viewing of the future and wrote down the fearful details.

Let's look at some of these insights! In her predictions she used the year 1926 as a base and continued from there. I don't know why, but that's what I read. By that time, the following will occur in the next 25 years.

186

For those who live the century through - in fear and trembling this shall do. "Flee to the mountains and the dens -to bog and forest and wild fens. For storms will rage and oceans roar when Gabriel stands on sea and shore, and as he blows his wondrous horn old worlds die and new be born.

The catastrophe above will happen just before the year 2026, and she goes on to explain just how the earth dies and is reborn. Let me go over the dies part. The reborn part you can find in some of the 2012 novels.

God's messenger from the heavens (comet) arrives and a great sound is heard as it passes through Earth's atmosphere and impacts Earth.
It causes wild storms and raging seas. A fiery dragon will cross the sky
six times before the earth shall die. Mankind will tremble and frightened be for the six heralds in this prophecy.

A comet' tail could, very well be considered a "dragon tail". The prediction could mean six major meteor strikes will occur which are spawned by the comet strike. It could also mean that before a terrible comet strike, we one earth will witness the huge ball of material getting closer and closer for 6 days before the eventual impact. Either way there is some bad stuff coming.

For seven days and seven nights man will watch this awesome sight. The tides will rise beyond their ken. To bite away the shores and then mountains will begin to roar and earthquakes split the plain to shore. And flooding waters rushing in will flood the lands with such a din that mankind cowers in muddy fen and snarls about his fellow men. He bares his teeth and fights and kills and secrets food in secret

187

hills and ugly in his fear, he lies to kill marauders, thieves and spies.

This comet strike evidently happens over a seven-day period or on the 7th day. One of the major issues of this event is worldwide flooding. Like the web-bot's prediction of Florida and other low-lying places will be underwater around the year 2009 and Dr. Casey's prediction that soon, the two halves of the United States will be split apart, these words are ominous and people begin killing each other over food.

The world upside down shall be, and gold found at the root of a tree. Yet greater sign there be to see as man nears latter century. Three sleeping mountains gather breath- spew out mud, ice and earth and earthquakes swallow town and town.
Not every soul on earth will die, as the dragon's tail [Comet] goes sweeping by, not every land on earth will sink, but these will wallow in stench and stink of rotting bodies of beast and man, of vegetation crisped on land.

This seems to be saying that the Earth axis will shift either before or after the comet strike. It is not known when. Her centuries ended on the 26th year after a normal turn of the century, so she was talking about the time between 2001 and 2026. Earthquake and Volcanic action becomes significant. Both could be caused by the comet or by the upcoming Earth axis shift.

Then shall be the Son of Man, having a fierce beast in his arms, whose kingdom is the land of the moon, which is dreaded throughout the world. Man flees in terror from the flood and kills, and rapes and lies in blood and spilling blood by mankind's hands will stain and bitter many lands. And when the dragon's tail [comet] is gone man forgets and smiles and carries on. To apply himself too late, too late for mankind has earned deserved fate.

After the comet, blood is spilled by war. Oddly, this includes war in space. Now, I have no idea where Mother Shipton could have gotten the space angle unless she actually saw it. The reference to the moon seems to be referencing involvement with others from outside the earth. The son of man in this verse is not a reference to God's son, but is, evidently, the leader of the "Christian Nations". He has a secret alliance with this beast thing. By using this alliance, he begins to take power away from the Moslem horde. Apparently, this beast he has in his arms, we can imagine that this has to do with the Biblical "mark of the beast" that will BECOME dread throughout the world.

Moslem War

War will follow with the work in the land of the Pagan and Turk the lily [USA?] shall be moved against the seed of the lion [Persia], and shall stand on one side of the country with a number of ships. <u>With a number</u> shall he pass many waters and shall come to the land of the lion [Persia], looking for help from the beast of his country <u>The lily F.K. shall lose his crown</u>, and therewith be crowned the Son of Man K.W. and the fourth year shall be preferred.

This seems to indicate that the Turks will eventually follow the Moslem nations in this pre-Tribulation War. "Lily to the rescue": the Moslem lion is beaten back by a nation with a strong Navy, possibly USA. Then some group goes a great distance across the water, probably the USA, to fight the Moslem Lion. Unknown to them, they will be aided by the beast or what Nostradamus called the Prince of Hell. Evidently the United States "Lily" and the Son of Man

189

"Prince of Hell" rule the world together for a time, but the old Prince takes full control after 4 years. Shipton even told us something about them with the peculiar initials. If someone becomes president with the initials FK I'm going to start worrying. Mother Shipton went on and on about how the world would be, not because she simply knew it or dreamed about it. She actually saw what was going to happen, by all accounts. Naturally she could not understand all of what she saw, so there are limitations to what we can get from the sightings, but it is fairly apparent that all that will happen has already happened "somewhere".

A Crown Explanation

Noted Near death researcher, Dr. Schofield, came up with a very good description of what or who the people are that "help us" whenever we get into the highest level of consciousness. He described it in 2 dimensions so even I could understand. In his explanation, he presented people that lived in a 2-dimension world. Someone from a 3 dimensional world tries to show them a sphere. It was believed that the sphere could allow a better communication between the two-dimensional beings, it could expand the awareness of the 2 dimensional guys, and it would explain the greater reality. For the two-dimensional being there was simply no way to view the sphere. In their world, it always looked like a circle. No matter how much the 3 dimensional people tried to explain the sphere, the more confused the 2-dimension people got, but the 3 dimensional beings kept on trying to enhance to advance in personal understanding, by means of communication and providing a feeling of joy to the 2-dimensional people. After a while, some understanding of the unbelievable could take place.

Vibration and Conscious Beings

It would seem that there are people that are sort of suspended in some type of in-between-land that can be seen and communicated with whenever one is vibrated to the Crown Consciousness level. Unfortunately, I cannot give you a very good explanation of this. My very dear friend, Ed Kaprock [This isn't his real name so don't even try to pry it out of me.] however, has been able to achieve this level of consciousness many times and believes that the people we see in this state can be considered to be teachers. Here is the problem. The teachers try to explain the things that you cannot understand at this level of understanding because you try to relate everything to our normal sensations excited by the workings of our universe. He tells me that over many "trips" to communicate with these people have allowed him to see things in a different light. He now feels that he is much more in tune with the consciousness around us and that that he cannot completely explain the difference. That brings me to crying.

Spirit Dimension

OK! You plowed through <u>"Life"</u> as a dimension, became cautious with the whole concept of <u>"Consciousness"</u> being a dimension and required for the universe to exist, but now I have just gone too far with something that is pretty weird. [OK! A lot of this is weird; but a Spirit gave this weirdness to us.] I won't be able to present a very good level of explanation for this very confusing dimension, because it has to do with soul transfer between universes and I can't find anyone to give me concrete answers about how it all works. Instead, I am going to give you general descriptions of the spirit dimension.

Make no mistake, the spirit is a dimensional quality required part for the life, death and existence of those who affect our universe, so do not simply ignore it.

As you recall, the notional equations were given towards the beginning of this book so you could see how all dimensional elements of our reality have a level of similarity and interaction. It must be that way or they would not be all part of our universal building blocks. From the equations we can see that we gain more spiritual energy as we increase our vibrational frequency and the spirit transfer energy is increased. While these make sense, it is important to see that if dimensional elements have a basis, the equations identifying their makeup should show a similarity to others identifying portions of the same universe. While this last dimension is very strange, it looks like the form of its energy profile is similar to the others.

193

Well, let me tell you this as a truth. Trying to put definitions around this "thing" is difficult. It has no mass and no motion so it isn't associated with either the nucleatic or photonic connective dimensions. That doesn't mean that there is no real connection element. This one may be the most important connection dimension we will ever understand. What I call the "Spirit" is the connective component of the life dynamo. It is, sort of, a key or gateway for living things between this universe and another. According to the Biblical testimony, the "LIGHT" was brought into the world ages before the sun became a beacon in the sky. Many claimed foul and say the Bible is a lie as light without a sun makes no sense. Today we know that light with a sun and no cognizant viewer makes no sense. Moses' writings here were possibly the very first introduction into Anthropics science ever recorded. This "light", as we later find out is the transfer key to a linked universe. Just like going through a black hole if you are matter or a magnetic monopole if you are and energy, the spirit or "light" dimension transfers through the released soul.

The Genesis story continues as this "light" was taken away from people before the end of the Pleistocene extinction and worldwide flood. Evidently, this "light" was in the early {Chosen People} survivors. Later it vanished again [presumable from intermarriage with people not having this "Light". Not to fear, God incarnate came to the earth and he left something called the "Holy Spirit" which was also interpreted as this "light". Now for the strange thing; this new "spirit" can and does become part of a person to allow for life transfer of a "released soul. Certainly the soul or consciousness is already living in a person, but with this "spirit", one could or can enter into a new universe called "Heaven". It was and is an extension of our life-consciousness beyond our death. Let me provide you with a few of the references that can be found about this mysterious "Light" and then we will discuss how or why we must consider it as a major dimension of our universe. Remember you can substitute spirit for light if you want to. The early Jews had no word for this thing so "Light" had to do.

"Jubilees"-The book of Jubilees is still considered a canon book of the Bible by some Christian sects. It contains many similar stories to our current Bible including stories about a strange thing called the "Light". *2:9- Nor may we take revenge on him because he has stripped us of the "light" [remember to substitute spirit]. He [God] marked out the borders of the world and created man in his own image with whom he hopes again to populate heaven, with pure souls.* [Not only note that without this light thing, the angels could not take vengeance on any of the heavenly host. They lost some substantial power. Also note that the word "again" is put in the verse to let us know that man was here before the Heaven war and was recreated after it was over, but that is another story. It seems that those who want to go to heaven will need this "Light" thing. It is some type of key.]

Please notice that this light [spirit] was required to allow SOULS to enter into Heaven.

I John-In the New Testament, the "light" continued as a connection between this world and the next. In this case, the word light had no sunlight meaning. It was talking about the way to heaven. ***Chapter 1:5-7****- This is the message we heard from Jesus and now declare to you: God is light, and there is no darkness in him at all. So we are lying if we say we have fellowship with God but go on living in spiritual darkness; we are not practicing the truth. But if we are living in the light, as God is in the light, then we have*

195

fellowship with each other, and the blood of Jesus, his Son, cleanses us from all sin. [The light is synonymous with the Holy Spirit.]

"Origin of the World"-This Gnostic Book provides the same type of crazy "Light" thing. *When Sophia [God] let fall a droplet of light. It flowed onto the water, and immediately a human being appeared, being androgynous- whom the Greeks call Hermaphrodites,* [Adam and Eve were made as one and were split apart as indicated in the Biblical record. Notice the light is discussed again.]

"Apocryphon of John"-A second Gnostic work says the same thing. *And he said, 'Come, let us create a man according to the image of God and according to our likeness, that his image may become a light for us. But the Epinoia [holy Spirit] of the light which was in him, she is the one who was to awaken his thinking. The Epinoia [Holy Spirit] of the light hid herself in him (Adamic man). And the chief Archon wanted to bring her out of his rib.* [This seems to indicate that the Archon rebels could not regain the light. This is a sad story, that is beyond this book.]

"Mishaf Resh"-This verse comes from the Zoroastrian religion. This verse about the strange "Light" thing comes from their version of their Holy Bible called Mishaf Resh.*Before the creation of heaven and earth Ali dwelt upon the sea. Then Ali went up to Heaven and solidified it. Out of His essence and light he made six gods.* [Like most Jewish Histories the "light" is a special component of life.]

"Zadspram"-Actually, the Zoroastrian religion, which uses this quasi-Bible, began around 700 BC, so it's been around a while. In this work the "Light" thing seems to address the same thing as the others. *From the seed which was the ox's, they would carry off from it and the "light" was entrusted to*

196

the angel of the moon in a special place, the seed was thoroughly purified by the "light" and was restored in its many qualities. [This seems to be the same "light"{key to heaven} referenced in the Bible]

"Emerald Texts"-Presumably written by King Thoth, who indicated that he had come from an Island nation known as Atlantis, this work provides a familiar theme. *The Children of Light dwelt among us* [Please note the "Light" description as angels had this "Light" thing unless they purposefully threw it away. ---*the children of light are different when they are not incarnate in a physical body". 32 were there of the children of the sons of light who had come among men.* [According to this book, only 32 of the sons of this "light" thing survived the flood. Jewish history suggested on 8.]

Maori Tradition-According to the Maori tradition, *the sons of Rangi and Papa were not unanimous in the decision to separate their parents [split the heaven and Earth apart] so a **huge war of the gods** followed the separation. After a 2ⁿᵈ war in heaven, Tane forced rebels to other worlds of darkness and despair. Tane forced the sky away from the Earth and Light came into existence.* [Notice again that light had nothing to do with the sun.]

While these don't help too much in our search for the details concerning our last dimension, the New Testament may help a little more.

New Testament Spirit

Typically, in the New Testament of our Bible, we find more pointed information about this light than provided in the book of "John I". Instead of being called "Light" this transfer miracle is called the Holy Spirit. Jesus indicated that he left the Holy Spirit to "fill" those on the earth. Bringing the Holy Spirit into one's self allows he or she to venture into heaven once a person has died. Without it there can be no admittance or transfer. What does that sound like to you?? It's the same thing as this "Light" that is being talked about by all these other writers. They simply didn't know to call it the Holy Spirit dimension. I know you are going to say, we lived in this universe for thousands of years without the Holy Spirit. How can it be a dimension? The answer is simple but I hate to say it…. Oh well.

The Holy Spirit was compactified before that time.---

Not really.

The Light or the Holy Spirit or whatever you want to call it has been an active dimensional element of our universe since it began. That being said, there was no knowledge of its importance to dimensionally set people as a means to traverse this universe and live in another and that brings us to being "absent from the body" as described in the Bible. Even if it doesn't bring us there, I'm going to discuss it so just read on.

Absent From The Body

Dying is really what gets the soul going. When it comes to dying, some people have very strong convictions. I know some of you are going to say the Bible tells you that the very instant you doesn't are absent from the body you [your souls] are present with the Lord, but the Bible may not exactly be saying that. Let's first look at Corinthians.

2 Corinthians 5:8- We are confident, I say, and willing rather to be absent from the body, and to be present with the Lord [It is only saying that dying isn't so bad, not that your soul zooms off to Heaven when you die. After all released souls may be very important to our "Reality".]

I know the thief on the cross was promised that *the day he died*, he would be with Jesus in paradise, but that does not necessarily mean that his soul immediately left this world and went to heaven. Remember that Jesus didn't even go to heaven for days after he died. If you remember in the Old Testament when Saul asked the witch of Endor to let him talk to Samuel, that she conjured away and the dead prophet, Samuel, woke up to tell Saul that he and both his sons would be slaughtered in an upcoming battle. Samuel's soul was not in heaven. In the New Testament many instances of demons were discussed where these demons could possess bodies and they were made to leave by Jesus and others. From what we can imagine, the countless people, in the New Testament, who were brought back to life when they had died had never ventured into Heaven and could tell no one what had happened to them. When a boy's dead body touched the bones of Elisha, he came back to life and knew nothing about any heavenly or hellish place. On and on we could go to show that when people die, they do not immediately go into heaven. In fact, in the book of Revelation we are told that the "Dead in Christ" will rise out of their graves in the last days. The reason they can do this is that they are still on the earth. The reason I'm saying all these things is not to make this a Bible class, but to instill the possibility that when you die, your spirit my go to our linked universe, but your soul does not. For what it is worth, let me tell you a possibility concerning the end of days. God incarnate [Jesus} will come back to this world to get his followers. Some are dead and some will be alive, but the souls of all of them will go to Heaven on that day. The method of transfer is what we can call the Spirit dimension.

The only reason I'm bringing up instances from Biblical texts is that you can relate to them. When you die, if you soul left this universe, there would have to be another soul entering the universe to allow for the transfer and keep both worlds "Neutral". I know you are going to say a baby gets born, so a soul would have to be transferred by means of the SPIRIT dimension. As the Bible told us souls are not leaving our universe, something else must happen. We can believe a number of released souls are free and either sleeping or active for short periods. Vibrating at the speed of light, time has no meaning to a released soul dimensional quality. If a new Life/self [Baby] is to be established, many times, released souls would become these now carnal entities just to make this dynamo even more confusing.

Reincarnation

Here is the supposition. These dead guys get tired of being dead and come back. By all accounts, almost always these entries are into babies. The kicker is that many remember something from past lives. Reincarnation is the attachment of a soul or "consciousness" to various living people over time. I know you have heard about Buddhists who believe that people could come back as animals, but that makes no sense. What would our consciousness do in an animal???? Some point to the Bible to gain assurance that reincarnation is an impossible task, but instead, the Bible does identify reincarnation "sort of"----

Hebrews 9:27 -"*For it is appointed for men to die once and after this comes judgment.*" [Some indicate that this indicates that a consciousness only lives once because of this verse. It doesn't say that. OK! It does sound like that, but other places seem to point towards a more loose characterization of the Spirit and a released soul.]

John 3:13-<u>*No one has ever gone into heaven*</u> *except the one who came from heaven--the Son of Man.* [Please understand what this is saying. I know it is not what some have told you, but <u>NO souls</u> besides those of residents called angels are in heaven. As many have died, either there is a huge pile of released souls or we live more than one carnal adventure.]

Matthew 11,14 - "*And if you are willing to accept it, John the Baptist is the Elijah who was to come.*" ---"*But I tell you, Elijah has already come, and they did not recognize him, but have done to him everything they wished. In the same way* *the Son of Man is going to suffer at their hands."* [Certainly if John the Baptist was Elijah, there is something going on that is reincarnation-like.]

Malachi (3:1; 4:5-6)- *"See, I will send you the prophet Elijah before that great and dreadful day of the Lord comes."*

John 9:2 -*"Rabbi, who sinned, this man or his parents, that he was born blind?"* [It is obvious that the first option (the man was born blind because of his sin) implies that he could sin only in a previous life.]

1 Corinthians 15:42,43,51,52- *Our bodies will be raised "in incorruption, glory and power"--- We shall not all sleep, but we shall all be changed. In a moment, in the twinkling of an eye, at the last trump: for the trumpet shall sound, and the dead shall be raised incorruptible, and we shall be changed* [As I mentioned before, there would be no reason for dead rising if their consciousness had already left the earth. I know some have told you that a nasty body that had been in the ground for thousands of years is what is going to rise, but that makes no sense. Also there is a reason why Paul indicated, "Not all sleep." That, simply, is that most people "sleep" after they die. Some sleep for many years. People don't go to heaven until the end of days. While they are on earth we can imagine that they either sleep or go to other bodies. Just how many billions of people have been on the earth from 40 thousand years ago until now anyway?]

1Thessalonians 4:15-17- *For the Lord himself will come down from heaven, with a loud command, with the voice of the archangel and with the trumpet call of God, and the dead in Christ will rise first. After that, we who are still alive and are left will be caught up together with them in the clouds to meet the Lord in the air. And so we will be with the*

Lord forever. [Again the dead people don't leave the earth until the last days so they are somewhere on the earth right now if all of them were ghosts, No one would want to buy a house or sleep with a light off or anything like that. Everything would be haunted by now if it weren't for reincarnation.]

There simply aren't enough Ghosts around to NOT have reincarnation.

Prerequisite for Reincarnation

You might ask, "Which consciousnesses reenter a body and which ones simply sleep like the prophet Samuel did?" I don't know the answer to that one. Some say that our creator wants your consciousness to continuously learn things and only after it experiences the things that are required or desired, "it" continues into additional entities. Others say that a consciousness emerges in another body if the previous person had to make restitution for something. Another indicates that it is like a circus. Around and around we go until God calls us home. There are probably a dozen others that you have heard about as well, and I can't tell you what may or may not trigger a reentrance of a consciousness into a new body. There are 4 things that I think I can say about this factor.

- **Death** is not the end.
- **When we die**, our consciousness is released so that it is possible to enter a brand new body.
- **When a body dies** our consciousness or more precisely our soul and spirit may be free to enter an already living body. I assume we would transfer into a young body, which may sound like a baby is not a living soul when it first is born. I'm not getting into that one, because there

potentially are more people living today that have ever lived before on the earth so a new born could very well have his soul at conception.

- **When God returns**, those consciousnesses that have linked with the Holy Ghost or "Spirit" [dimensional transfer agent] will transfer to the other universe. Souls without this "spirit" link cannot.

You might ask is there any proof of reincarnation besides some insecure person not wanting to have to think about death being the end of his existence. The answer is there is.

Reincarnation Proof

While there are many, many reports of reincarnated people, Dr. Stevenson methodically documented the child's statements of a previous life gotten from his encounters. After getting details he identified the deceased person the child remembered being, and verified the facts of the deceased person's life that match the child's memory. He even matched birthmarks and birth defects to wounds and scars on the deceased, verified by medical records. His strict methods systematically ruled out all possible "normal" explanations for child's memories that didn't make sense.

Dr. Stevenson devoted well over forty years to the scientific documentation of past life memories of children from all over the world. He collected over 3000 cases in his files alone. Many people, including skeptics and scholars, agree that these cases offer the best evidence yet for reincarnation. Children remembered their past lives. Under hypnosis, others regained knowledge that they could not have had. Some were able to speak in languages they had never learned. On and on and on and on we could go. It is very difficult not to accept this huge database of information that confirms the reincarnation probability.

Not Reincarnated

Like Samuel in the Old Testament, it reasonable to believe that many consciousnesses do not become reincarnated and stay in some limbo where they can be teachers while people dream or leave there bodies on purpose or leave there bodies to the trauma associated with near death experiences. Certainly some consciousnesses not associated with living people are sleeping until the time they will be raised and go to another universe and a place called Heaven. Some of the consciousnesses are good and some are not. One thing that is important to remember; all of these consciousnesses are REQUIRED for our universe to exist. We should consider that the consciousnesses [souls and spirits] are both all over the place and ready to get us going.

Conclusions

Hopefully this has peaked your interest in what our universe ACTUALLY is. If you tell me it is a length, width, and depth, I'll shoot myself. I am sorry for the Biblical instruction, but I did not know how to describe these important dimensions in another way that could be easily interpreted. Here are a few of the things I hope you got out of this book.

- All matter is made up of vibrations not particles.
- Beyond that, everything in the universe can be broken down into vibrations. This includes not only the things we can see like planets and tiny little pebbles, but also electricity, magnetism, Photons, Nucleons, the spirit, the soul and even life.
- Instead of three dimensions, there apparently are three dimensional dynamos, each made up of 3 dimensions to build our universe and a 4thb dynamo that activates the other three.
- The Structural Dynamo is similar to the dimensional elements we previously understood except that they are totally based on vibration rather than distribution.
- The Aetheric Dimension characterizes matter. As the energy increase the vibrational frequency increases.
- The increase in frequency attracts fermions together in a quantized fashion that insures fermionic regeneration along a close dimensional string.
- The Gravity dimension is perpendicular to the fermionic one. It is completely connected to the fermionic one and

advances gravity into all matter. Increasing frequency increases the effect of gravity. Gravity decreases by the square of the distance between it and its fermionic host.

- The Nucleatic dimension is a connective one. It brings matter together in accordance with the frequency base of the fermionic dimension.

- The nucleatic force can sometimes disappear as if it can transfer between 2 universes.

- The Operational Dynamo controls motion. It consists of Magnetic, Electric and photonic dimensions. Each is perpendicular and connected to the other.

- The electric dimension builds energy by displacement. It can be associated with potential energy. The higher the frequency, the more electric field is generated.

- The magnetic field is associated with the electric field the farther the magnetic field is from the electrical source, the energy level is reduce by the square of that distance.

- Photonic dimension makes vibrational patterns associated with light. Requiring both electrical and magnetic fields, energy in increased with the frequencies of the other 2 dimensions.

- Sometimes photons appear to disappear as if they are interface between universes and can transfer back and forth.

- The M-Theory with its "Super-Symmetry" insure that there has to be a connected universe for this one to exist.

- This connected universe must be similar to ours, but go backwards in time to insure there is no loss of energy, matter and life in our universe.

- Out-waves building particles leave our universe and electro-magnetic in-waves come in at the same rate to keep our universe stable. As an "out-wave" leaves our universe, it become an operational dynamo in the associated universe.

- The Life dimensional dynamo is the most difficult to understand it builds life in the universe.
- The self-dimension characterizes matter as living or not living the level of life is associated with the vibrational frequency of this dimension.
- Anthropic Science tells us this dynamo is the most important one when "defining" reality in our universe.
- Associated with life is consciousness. We spent a lot of time on this subject because it is difficult to understand the separation of consciousness and life.
- Continuousness in this sense is consciousness of the union of souls defining our existence. If we change the resonance of this consciousness, we change reality for a short time.
- The last dimension element of our universe besides time, is something I called the Spirit dimension. This dimension restricts and allows transfer of living entities and consciousness between universes. While it is fairly easy to define it is extremely difficult to understand simply because it has no physical characterization beyond the transfer quality. Its necessity is only understood when we die.
- Time is made up of forward and backwards time and a transfer dimension I called Lateral Time.
- Lateral time allows consciousness transfer between universes to allow people to witness the future and the past.
- You should have also gotten a better understanding about how the Christian religion fits with all of this stuff. How God incarnate came to reestablish the link between Heaven and Earth. How demons could be understood.

208

How reincarnation could be addressed. How the final rapture of the Church or the 2nd coming of God incarnate could fit into a world not of 4 dimensions as you once thought but in a world of 12 dimensions.

- This was not a religion book, but religion must coincide with our mathematical models or we have one of 2 problems. Either the science is wrong or the religion is wrong or incomplete.
- Also science must embrace the characteristics that thousands of people experience including NDEs, Out of body experiences, astral projections, seeing the future, and reincarnation. If your science can't explain or use these common elements of our life, the science is wrong or incomplete.
- Hopefully, no one finds that "Brown Note" again.
- I hope this book has been useful to you when trying to interpret what otherwise was considered anomalous in our world. If too many things don't fit, you must try to change your definitions. This is a new definition.

The End

About the Author

Steve Preston is a long time author of scientific, esoteric facts. His series on the creation of mankind is shown below. The series focuses on the painful truths rather than whitewashed details that make us comfortable. If you are interested in the truth instead of comfort, please continue to read and, while you are at it, review other works by Mr. Preston as shown below.

Development of Mankind

The First Creation of Man-book 1 History of mankind
The Second Creation of Man-book 2 History of mankind
The Creation of Adam and Eve-book 3 History of mankind
The Antediluvian War Years-book 4 History of mankind
Man After The Flood-book 5 History of mankind
A Closer Look at Ancient History-book 6 History of mankind
A New View of Modern History-book 7 History of mankind
The Twentieth Century and Beyond- Book 8 History of Mankind

Bible History, Correction, and Analysis

Abraham to Moses-First part of the Bible
Adam's First Wife-Story of Lilith
Adam to Abraham- Second Part of the Bible
Closer Look At Genesis- 200 ancient text confirm Genesis
Exploring Exodus- Reviewing the Details of "Exodus"
Errors in Understanding- Interpretations of the Bible
Expanded Genesis- Apocrypha and other Jewish texts
Exploring Genesis- Reviewing the details of "Genesis'

Incarnations of God- How often did God become Incarnated?
History Confirmed By The Bible- Science confirms the Bible
Moses Saved Egypt- How the Jews eliminated the Hyksos
Moses to Jesus- Third part of the Bible Series
Mysteries of the Exodus- Proofs of the Exodus
New look at the Bible- Questions in Interpretation
Old Testament Used By Jesus- Ancient Jewish texts
Understanding the New Testament-4th part of the Bible Series
Why the King James Bible Failed- Issues with KJB

Ancient Technology and Life

Anakim Gods- History of the Ancient Giant/gods
Ancient History of Flying- Ancient flying
Kingdoms Before the Flood- Pleistocene humans
Living on Venus- Venus before the Pleistocene Extinction
Martians- Ancient Life on Mars
Mysterious Pyramids- Who made the Pyramids?
Victory of the Earth- History of our Earth
Not from Space- UFOs are not from space.
Amazing Technology- Descriptions of prehistoric capabilities

Ancient and Modern War

America's Civil War Lie- Truth about the Civil War years
Behind the Tower of Babel- Story of the Bharata War
Driven Underground- Fear in the Bharata War
Four Armageddons- The 4 major wars that destroyed mankind
Six Deaths of Man- Destructions of mankind
World War Before- The Pleistocene War
World War with Heaven- The Angel and Anak War
World War Zero-The Bharata War
When Giants Ruled the Earth- History of the Titan Giants
Sex Crazed Angels- What caused the Heaven War?

Current Events and Fears

Allah' God of the Moon- Terror of Muslims

American School Disaster- fear in our country

Can We Save America? - Fear in the USA

Scythians Conquer Ireland- A History of Ireland

Fast History of MILES Training- Laser based Army training

Great American Quiz- Unusual details of American History

Make Your Own Global Warming

Truth About Phoenicia- The Evidence -First in America

Monsters are Alive- Post Pleistocene Monsters

Promote the General Welfare- Fear in USA

Our Very Odd Presidents- President review

Terror of Global Warming- Fake issue uncovered

The Antichrist- Many demonic possessed rulers

The Bad Side of Lincoln- Negative side of a great man

The Devil- Of Demons and their master

Vampires among Us- How Demons and Vampires are similar

Humans on Display- Slavery and Human Zoos

New Look at Physics

Amazing Technology- Pleistocene Technology

Anthropic Reality- We control our Reality

Consensus Science- Fake Science

Complex Earth- Truth behind Earth's development

Is Time Travel Possible? Science of Time Travel

Retiming the Earth- Eliminate of Nuclear Decay Errors

Releasing Your Consciousness- Beyond our SELF

Slip Through a Wall- How to walk through solids

***Our 12-Dimensional Universe**- New science of our Universe*

Mystery of Photons and Light- Science of Photons

Of Heaven and Hell- scientific descriptions

Meaning of Life and Light*- Detains of New Science*
Vibrational Matter*- New Science of Quantum Fluctuations*

New Look at Biology
DNA of Our Ancestors- Tracing DNA of ancient man
God Didn't Make The Ape- New science on ape Evolution
Lizard People- Mutated People of the Bharata War
Creation and Death of Dinosaurs- Why Dinosaurs died
Races of Men- Tracing DNA of Humans
Tracing Cro-Magnon to Jesus-
Self, Soul, Spirit- Three components of Life
Self-Virtualization- New science of reality
True Happiness- Self Actualism and Beyond
Life Resonance- Unusual capabilities of men
Awaken the Departed- We can talk to the Dead
Biophotonics and Healing- How Photonics used in medicine

www.ingramcontent.com/pod-product-compliance
Lightning Source LLC
Chambersburg PA
CBHW071422180526
45170CB00001B/187